Medical Ethics, Ordinary Concepts and Ordinary Lives

Medical Ethics, Ordinary Concepts and Ordinary Lives

Christopher Cowley

palgrave
macmillan

© Christopher Cowley 2008

First published 2008 by
PALGRAVE MACMILLAN
Houndmills, Basingstoke, Hampshire RG21 6XS and
175 Fifth Avenue, New York, N.Y. 10010
Companies and representatives throughout the world

PALGRAVE MACMILLAN is the global academic imprint of the Palgrave Macmillan division of St. Martin's Press, LLC and of Palgrave Macmillan Ltd. Macmillan® is a registered trademark in the United States, United Kingdom and other countries. Palgrave is a registered trademark in the European Union and other countries.

ISBN-13: 978–0–230–50690–9   hardback
ISBN-10: 0–230–50690–9   hardback

This book is printed on paper suitable for recycling and made from fully managed and sustained forest sources. Logging, pulping and manufacturing processes are expected to conform to the environmental regulations of the country of origin.

A catalogue record for this book is available from the British Library.

A catalog record for this book is available from the Library of Congress.

10   9   8   7   6   5   4   3   2   1
17   16   15   14   13   12   11   10   09   08

Printed and bound in Great Britain by
CPI Antony Rowe, Chippenham and Eastbourne

Elišce a Františce

Table of Contents

Acknowledgements

A number of chapters include extracts of material previously published, and I would hereby like to thank each of the following journals for their kind permission to reproduce then. In most cases, however, the material has been heavily revised, not only to fit in with the new context, but sometimes also in an attempt to improve it.

- Chapter 1 features extracts from 'The dangers of medical ethics', in *Journal of Medical Ethics*, vol. 31, no. 12, December 2005.

- Chapter 2 features extracts from two articles, 'Changing one's mind in ethics', in *Ethical Theory and Moral Practice*, vol. 8, no. 3, June 2005, and 'Why philosophers shouldn't teach medical ethics', in *Discourse*, December 2005.

- Chapter 6 features extracts from two articles, 'Narrative and the personal in ethics', in *Practical Philosophy*, vol. 8, no. 1, Summer 2006, and 'The last chapter in the story: A place for Aristotle's *eudaemonia* in the lives of the terminally ill', in *Online Journal of Health Ethics*, May 2006.

- Chapter 7 includes extracts from two articles, 'Suicide is neither rational nor irrational', in *Ethical Theory and Moral Practice*, vol. 9, no. 5, November 2006, and 'The Diane Pretty case, and the occasional impotence of justification in ethics', in *Ethical Perspectives*, vol. 11, no. 4, 2005.

- Chapter 10 includes an extract from 'In praise of fudge: Euthanasia and the law', in *Richmond Journal of Philosophy*, vol. 10, Summer 2005.

I would here like to express my deep gratitude to my two greatest philosophical teachers, Carolyn Wilde (my doctoral supervisor at Bristol) and Marina Barabas, both of whom I have been glad to have remained friends with. This book comprises too many of their ideas to list individually, even if there is always a risk that I have misinterpreted them.

With regard to this manuscript, my greatest debt is to David Cockburn, who read almost half of it and provided very useful feedback. My thanks also to the School of Philosophy at the University of East Anglia for important feedback on some of the material, and especially to Gareth Jones.

This book is aimed not simply at philosophers and philosophy students, but also at health care professionals. Over the past four years I have been careful to discuss many of the ideas with friends and colleagues from the medical world, some of whom provided useful feedback not only on points of clinical detail and procedure, but more importantly on the very nature of the medical world itself. I am especially grateful to Mark Wilkinson, consultant pathologist, for lengthy conversations, but also (in no particular order) to Rosalyn Proops and Richard Beach, consultant paediatricians; Maggie Wright, consultant anaesthetist; Leslie Bowker, consultant geriatrician; Mike Laurence, Christopher Hand and Janie Anderson, general practitioners; Hugo De Waal, consultant psychiatrist; and Stephen Domek, medical student.

Introduction

This is a book about medical ethics. However, it is not a textbook focussed on the main *issues* in medical ethics; nor is it a textbook focussed on the main ethical *theories* that can be brought to bear on such issues; and it makes no claim to be comprehensive. Where a particular issue, such as abortion, has been extensively discussed elsewhere, this book will not be recapitulating the arguments for and against in any systematic detail. Instead, this book will have two related aims, one critical and one positive. The critical aim, mainly in Part I but also throughout, will be to challenge certain key assumptions underlying the very framework of most discussions in mainstream books, textbooks and academic journals. I accept the risk of using a single word, 'mainstream', to describe many different opponents, but I do believe they share enough assumptions about ethical discussion for me to mount a general criticism and avoid a 'straw man' fallacy.

I claim, that in an effort to deal with the difficult problems that characterise the world of medicine, mainstream philosophers have arbitrarily omitted many of the ethically relevant features of those problems in order to reduce them to more tractable technical puzzles. The most gratuitous omissions have been the patient's distinctive point of view on the problem, the patient's ordinary life that provides the wider context for his point of view, and the ordinary language and concepts by which the patient makes sense of the problem. The result is often a confused, oversimplified and misplaced understanding of the problem, and a naïve faith that with enough ingenuity a solution can always be found.

Part II then proceeds to examine a number of issues associated with pregnancy, birth and life with a view towards remedying the above omission. But sometimes it does not confront these issues head-on. Instead, there is a good deal of preparatory work involved in first understanding the place of normal pregnancy, birth and childhood in our lives. Only in this way can we hope to understand the wider meaning of something like abortion (to which I turn in Chapter 5). Similarly, Part III is primarily interested in the place that old age and death has in our lives, both in the sense of the old person's experience and in our relations with the elderly. Only by considering this wider meaning of old age and death can we begin to make sense of the debates surrounding suicide and euthanasia.

Because of this emphasis on ordinary lives, the discussion will not be limited to medical examples, but will often stray into examples from ordinary life in order to draw out a comparison or contrast. After all, the meaning of the very concepts used in discussions of medical ethical issues is primarily located in our *non-medical* experience, and originally learnt as children in familiar contexts such as families, schools and churches. As such, one of the main themes of this book is an interest in the meaning of concepts in the wider sense, a sense that includes the way that people spontaneously use the concept in ordinary thought and speech, and that includes what I call the 'resonances' of the concept: the presuppositions and implications and connotations behind its use, the other concepts to which it is logically related, its particular application by a particular human being existing at a particular point in his life. Although I use the word 'ordinary' twice in the book's title, this should be taken as a contrastive term rather than as a specific technical term. By an ordinary person, I mean someone who does not work in the medical world, but instead works in an office, or a factory, or out in the fields. One of the things I will be claiming is that the medical world is utterly unique and extraordinary. In addition, an ordinary person, leading an ordinary life, should be contrasted with the hapless heroes of the contrived schematic examples that philosophers sometimes use to make their point. Finally, by ordinary language, I mean non-technical language and especially non-philosophical language: the stuff of the hundreds of conversations we have during the day on subjects about which I or my interlocutor is not an expert.

I have been assuming throughout this book that my reader is the same type of regular reader of the *Journal of Medical Ethics* (*JME*) and of the *Hastings Center Report*. I think the issues raised are important and relevant for a larger group than just academic philosophers, and yet I do not have the space to introduce the basic debates in normative and applied ethics in textbook fashion. The *JME* reader will be familiar with the specific problems that constitute the new discipline of medical ethics, with the main arguments for and against a practice such as abortion and with the broad outlines of the major public policy solutions to such problems in the United Kingdom. Although my knowledge is mostly of the British environment, most of what I say will be of relevance to any Western country.

Because of this wider target audience, I have opted for a more informal prose style and have tried to avoid cumbersome technical vocabulary as much as possible. This is not just a point about readership, however; there are also sound philosophical reasons for my suspicion of technical vocabulary in moral philosophy, and these reasons are directly relevant to the central arguments of the book. I lay out those reasons in Chapter 1.

Alongside the interest in ordinary lives and ordinary language, and the emphasis on the wider meaning of concepts, there are a number of other interrelated themes in the book that are worth introducing here. In many places I distinguish between two kinds of knowledge, which I sometimes call 'experiential' and 'propositional'. A classic case of propositional knowledge is that 'all human beings are mortal', whereas many people lack the experiential knowledge gained from living through the dying of a loved one. I contend that many discussions in medical ethics take place at the level of propositional knowledge between participants who lack sufficient experiential knowledge, and that this impoverishes and distorts the ensuing discussions.

Most problems of medical ethics are actually threefold: there is an *impersonal* (or public) problem, there is the *policy* problem and there is a *personal* problem. The impersonal problem is that of trying to find out whether abortion in a given type of situation, or for a given type of reason, is right or wrong for *all* people who find themselves in that type of situation or who request an abortion for that type of reason (the solution is therefore 'universalisable' in classic Kantian fashion). Typically, this impersonal approach is driven by the urge to build up a coherent theory for all types of situations and issues. Second, the policy problem involves the search for a pragmatic solution, once we accept that the public problem is not open to a universally satisfactory solution; as such the policy solution will necessarily be some sort of compromise that will satisfy as many and offend as few as possible. In principle, the policy problem is still informed by the impersonal public problem: the ideal policy will be one acceptable to all parties, once they have solved the impersonal problem. Third and finally, there is the personal problem of what an individual person is to do, given the legislative background and the social attitudes.

In this book I am much more interested in the personal problem, in how the individual makes sense of the problem (what precise words he uses to describe it) and of the options available for its resolution (which options *he* considers viable), and how the individual then deals with the consequences of the option he has chosen, and how he sees himself as a result. As will be obvious, this can get very complicated and unwieldy, but ... such is life, and the simplifying theories ignore such complexity at their peril.

Although my sympathies will become obvious, I find it difficult to say that I 'support' or 'do not support' a given policy, and I certainly refrain from declaring myself aligned for or against any particular ethical theory. To some this will seem to be 'sitting on the fence'; but given the shrillness of those engaged in many of the issues I will be considering, there is still something to be said for a more humble, hesitant bit of fencesitting in the larger discussion. The mainstream philosophers whom I will be criticising are much more useful in generating coherent policy, and that might be a drawback to my more laboured, nuanced approach. But I would like to think that there is room for a *subsequent* policy discussion, once the insights from my approach are taken on board – I have not attempted such a discussion here, however.

My approach has sympathy for and a lot in common with two trends in ethical theory, but it is worth distinguishing it from them at the outset. First, ethical intuitionism, now less popular than it once was in a times of Moore and Ross (the modern version is defended by David McNaughton, among others), involves the thought that we have direct, quasi-perceptual access to the truth about, say, what ought to be done in a given situation, without recourse to theory. The second trend is variously called anti-theory (Bernard Williams), or uncodifiability (John McDowell), or particularism (Jonathan Dancy). However, both trends, as I understand them, remain essentially *realist*; that is, they involve a substantial metaphysical commitment to an independent, discoverable, and most importantly *singular* realm of ethical facts about, say, what ought to be done. I wish to be agnostic about the metaphysics since I am more interested in the individual's own experience of the world and of the ethical claims he discovers therein. I have chosen not to engage with the intuitionist and particularist literature, in order to start 'from scratch', as it were, and build up a position that focusses more on questions of meaning than of metaphysics. What is crucial for my account, however, is that the meanings are very much objective, that is, they are discoverable, independently of the enquiring mind, and it is possible for that enquirer to be mistaken about them. In other words, it is the singularity of the realist realm that I will be implicitly challenging, not the objectivity; and my account should never be dismissed as merely subjectivist for this reason.

However, my approach *is* intuitionist insofar as I often rely on certain intuitions in the reader in order to make my point. All philosophers do this to a certain extent, but I may be accused of overdoing it and thereby opening myself to criticism from a reader who simply doesn't share that particular intuition. For example, I have described suicide in

Chapter 7 as inspiring a peculiar kind of horrified fascination – and this just might be my own morbid response and no one else's. However, given the nature of ethical disagreement as I discuss it in Chapter 2, this is a risk I have to take.

My approach to ethics is not very common, but has been directly inspired by a number of philosophers who have developed similar insights much more powerfully than I: Peter Winch, D. Z. Phillips, Cora Diamond, Stanley Cavell and Raimond Gaita. There are fewer writers in medical ethics who share this approach, but possibly the most well known are Carl Elliott, whose (1999) contained many useful insights for this book, and Anne Maclean, whose (1993) comprised a blistering attack on the utilitarianism of John Harris and Peter Singer. Finally, I should mention Stephen Mulhall, not normally interested in medical ethics, who wrote a wonderful review (2002) of Jeff McMahon's book (2002), on which I have drawn in my own criticism of McMahon in Chapter 1.

Part I A Critique of Mainstream Medical Ethics

This first part sets the stage and introduces the main criticisms that I wish to make of mainstream medical ethics, criticisms that I will be revisiting through the rest of the book. I consider two prominent examples of mainstream literature, but rather than attack their arguments directly, I examine their whole methodology and approach. My aim is to show that such an approach is inappropriate to the peculiar subject matter of ethics, and to capturing the full reality of complex ethical situations. In order to show something of the possible complexity of this reality, I then examine the structure of the ethical disagreement between a vegetarian and a carnivore about the ethical permissibility of eating meat. This disagreement is typically not resolved by persuasion but by some sort of conversion, where one can 'bring the other to see' that the action in question is right or wrong. If such complexity is distinctive of ethics, then it is even more distinctive of medical ethics, precisely because of the radical contrast between the medical world and the ordinary world.

Part I: A Critique of Mainstream Medical Ethics

1
Technical Language and Ordinary Language

In order to focus my critical discussion of the mainstream approach to medical ethics, I have selected two examples: Beauchamp and Childress's *Principles of Biomedical Ethics* (1979) and Jeff McMahon's *The Ethics of Killing* (2002). Both are faithful representatives of the field. *Principles of Biomedical Ethics* is still considered the main textbook on the subject, and its fifth edition was published in 2001. McMahon's book is an enormously detailed and rigorous analysis of the arguments for and against abortion and euthanasia, and was received with widespread critical acclaim. Rather than a systematic critique of the two books, furthermore, I have plucked only a single section of argument from each, a section that in each case, I believe, exemplifies their overall approach. While I am aware of the risks involved in taking such narrow selections, they are worth running because of the limitations of space and because of the depth to which I want to pursue my critique.

The section of the Beauchamp and Childress book for which they have become most famous is that concerning the Four Principles (which I will henceforth capitalise). These Principles – of beneficence, non-maleficence, autonomy and justice – have become very influential. Evidence for their influence can be seen in the publication by the *Journal of Medical Ethics* (*JME*) of a 'top ten' of the most-downloaded *JME* articles in 2006.[1] On that list the third, fourth and eighth, by Gillon, Beauchamp and Macklin respectively, all defended the Four Principles. So despite criticism from a variety of quarters, they are still going strong. In what follows I will take the Principles, and their application to the problems of medical ethics, to be broadly familiar and so will not describe or defend them in any detail here. Nor do I want to get into the main criticism that has been levelled elsewhere against the Principles: that there is no clear hierarchy among them to resolve those cases

3

where different Principles would advocate incompatible actions. Instead, I want to examine how these Principles – and indeed any ethical principles – are supposed to work in practice.

What is initially noteworthy about the first two Principles is that beneficence and maleficence are technical terms, that is, they are not used by the vast majority of people, even in situations of clear ethical import. Sometimes, within the context of an enquiry in an academic discipline, new terms will have to be invented in order to label new phenomena, achieve the necessary precision and make the necessary distinctions. While the rest of us describe water as clear and colourless, the chemist will also be interested in its pH, surface tension and solubility. Beauchamp and Childress's first two Principles do not seem to say much more than 'do good' and 'do no harm', and this encourages the suspicion that they were searching for a certain academic gravitas. As I hope to show, this is not a trivial point.

If the first Principle is indeed no more than an injunction to do good, it is then unclear how this is supposed to work without helping us to *identify* the good, or at least the most good. Surely most of us, most of the time, are trying to do good; it's just that we often have radically different ideas of what the good is. In the same way, most of us will agree that we shouldn't give our children too many sweets or too much telly. But without some way to determine the correct amount of sweets or telly, the general injunction does not achieve much. At this point the utilitarian usually steps in to offer to quantify the good, and thereby to provide a common currency by which to measure one available option against another, with the general injunction not just to do good but to maximise it. But this faces the familiar objection that the sheer variety of things we are inclined to call good cannot plausibly be reduced to a general notion of happiness or welfare. Certainly there is a limited place for a refined utilitarianism in the allocation of scarce resources, both in a public health care system and in a whole country. But even here the National Health Service (NHS) management will not be much helped by a Principle of beneficence.

What about 'non-maleficence', or the older Hippocratic oath? Neither version is going to enlighten anybody, that is, tell them something they didn't know before; and neither version is going to solve many ethical puzzles. After all, in most cases harm is *by definition* what I shouldn't do. If I understand a particular action as harmful, then that is already a reason not to do it. You and I may disagree about whether a proposed action is in fact harmful, or whether a proposed action has too high a risk of causing harm, but then we're into a different kind of disagreement, since we both agree that harm is something to avoid. Or you and I may agree that a particular action is harmful, but disagree about whether it is *justified* by the ensuing benefits, and this will be another type of disagreement. After all, every time I give an injection, I am inflicting deliberate harm.

If the Principle has merely led us on to these substantive and important questions about the reality or likelihood or justifiability of harm, then why not go straight to these questions, without the needless obfuscation by the Principle of non-maleficence? There might be two interrelated reasons. First, perhaps there are occasions where a *slogan* can remind us of an important consideration that we are otherwise at risk of neglecting. Imagine a doctor who is tempted to experiment with a new surgical technique, but decides to postpone it when he recalls the slogan and realises that he has not yet properly explored the risk and degree of harm that might be associated with the technique. Second, the Principles could act as 'a useful "checklist" approach to bioethics for those new to the field' (Harris 2003), to make sure all relevant angles are covered.

Neither reason holds water, however. First, the surgeon in our example is more likely to hesitate when an experienced colleague warns him that the technique is unsafe or when he sees the apprehension in the patient's eyes or when he reads an article about the increase in medical negligence suits. In each case, however, the warning is much more specific than the bland injunction to do no harm. The thought that a surgical technique might be harmful should occur spontaneously to anyone with authority to perform it; if it doesn't, it is hardly likely that a mere slogan can get through to them. Second, anybody who is ethically obtuse enough to need a checklist of principles would not be capable of interpersonal relationships of any complexity, and certainly should not be practising medicine.

What about the other two Principles? Certainly 'autonomy' has a solid philosophical pedigree in the work of Kant. And at least the term is a little more familiar, although hardly in common usage. We may take the notion of respecting another's autonomy to have two essential and interrelated meanings: first, to respect them as a moral equal;[2] and second, to respect them as a sovereign will, especially in matters concerning their own life, body and property. Once again, there is the problem that these two versions of the Principle are obvious and should have been understood by all of us by the age of five; if we do not yet understand it as adults, no Principle can help us. But as with the concept of harm, the Principle obscures rather than clarifies the real philosophical questions. What does it mean to treat another as a moral equal, especially, for example, when the other is a patient who is physically abusive to staff or who refuses to be treated by a nurse because he has the 'wrong' skin colour? Does respecting the other's sovereign will involve standing by while the other harms himself?

Finally, the Principle of justice. At least the other three Principles are not so vulnerable to cliché: which of the comic-book superheroes hasn't fought for justice? This concept is mired in all sorts of extra complexities invoked by the implicit reference to a community of other people with varying claims on limited resources. The classic problem here is to decide on the central criterion of a just allocation. Is it to be age, desert, wealth, short-term need, long-term need, benefit – or should everyone get an equal share? Is full transparency a prerequisite of justice? Should one strive for a result or a procedure which is fair? How binding are precedents? How consistent should the NHS be across the country, when local needs vary? Are higher taxes and free health care essential to a just society? Should it be doctors who worry about fair resource allocation or should it be managers, or politicians? The idea behind this bombardment of rhetorical questions is to underline the fact that justice is a *huge* issue in medical ethics and anywhere else. Of course everybody wants to be just, and have declared a concern for justice ever since they were able to shout 'that's not fair!' in the schoolyard. But disagreements between advocates of different distributions will be much more complex than the one-dimensional disagreements about the appropriate amount of sweets or television time for our children's good.

There is another interesting problem with Beauchamp and Childress's attempt to reduce all the other ethical concepts to these Four Principles. Imagine a homeless patient is about to be discharged and asks the doctor for some money for a bus fare and a cup of coffee. Is 'autonomy' the best word to use in this context, as opposed to more ordinary words like charity or generosity? Beauchamp and Childress might claim that charity and generosity could be captured by the Principle of beneficence, and that this was now the commanding Principle. But in reducing concepts like this, in fitting the round peg of charity into the square hole of beneficence, we are losing crucial nuances from the original complexity and thereby undermining whatever answer the theory comes up with. The worry is that in sacrificing the richness of ordinary ethical language for the alleged precision of technical ethical language, we are not so much getting at the root of the problem as describing a *different* problem altogether.

One of the defenders of the Four Principles, Ranaan Gillon, claims that 'ethics should be basically simple for it is there to be used by everyone, not just people with a PhD in philosophy or theology' (Gillon 2003). In terms of the injunction to avoid needless complexity (e.g. elaborate metaphysical constructions), I would certainly agree. The problem is that he confuses simplicity with lucidity. Philosophers should be interested above all in lucidity, and they can bring their distinctive argumentative and analytical skills and experience to bear on complicated ethical problems in order to make them more lucid, both to those with a Ph.D. in philosophy and to those without. Some *prima facie* difficult ethical problems might turn out to be simple and solvable; but striving for lucidity also involves an honest willingness to recognise irreducible complexity, and to do justice to that complexity in their analysis of the problem. Indeed, in striving for lucidity the philosopher may often make the original problem more complex than was first apparent.

One source of irreducible complexity that is insufficiently acknowledged by mainstream philosophers such as Beauchamp and Childress, I believe, is the wider meanings of the ordinary ethical concepts used by ordinary people to make sense of the problems they face. A narrow meaning of a concept is what you might find in a dictionary; a wider meaning involves the resonances associated with the concept and with its application. Technical terms like 'beneficence' are not supposed to have any connotations: they are just supposed to describe what is there. And this is considered an advantage. However, sometimes the connotations will be essential to the ethical reality of the situation, and in applying a technical term the philosopher may miss part of that reality.

In academic disciplines such as medicine and chemistry, and indeed in other areas of philosophy such as metaphysics, a new, technical language is essential to understand the new, unfamiliar problems which the discipline has discovered. Ethical theory, as an extension of metaphysics (since it is interested in the status of things like obligations), also has a legitimate claim to a technical vocabulary; so does what might be called ethical anthropology, the effort to describe the ethical practices of our own and other societies, present and past. But the sort of ethics that is the focus of authors such as Beauchamp and Childress is unlike any other academic discipline and unlike any other area of philosophy, precisely because its problems are essentially familiar to each of us, and ethical arguments are essentially accessible to each of us. As such, philosophical ethics is *answerable* to the real world of ordinary people and to the language they use. At the very least there is no need for further technical language; and at most the technical language will distort the familiar problem. As a result, it is particularly important for a work on medical ethics to consider not only problems that are typical of the world of medicine, but also problems from ordinary life, for the ordinary people who encounter the problems of medicine (as patient or doctor) will inevitably bring their ordinary ethical concepts and ordinary ethical language to bear on the medical world. It is possible that the medical problem will be of such complexity and unfamiliarity that they will then find it difficult to orient themselves, and this possibility should always be recognised, instead of reaching for the false sense of security offered by the new technical terms.

The above is a central theme of the book and will require further argument in support. For the moment, however, I want to consider my second example of a mainstream philosophical approach to ethics.

Fearless thinkers and monstrous thoughts

About two-thirds of the way through his book, Jeff McMahon makes the following remarkable admission, which is worth quoting at length:

> I expect that many readers will regard the fact that my position supports the permissibility of infanticide in certain conditions as a *reductio ad absurdum* of the position itself. But the implications of the view are in fact even more shocking to common sense than I have so far acknowledged. Let me cite the worst-case example. Suppose that a woman who wants to become a single parent becomes impregnated through artificial insemination, but dies during childbirth. She has no close friends and no family – no one to claim the child. The newborn infant is healthy and so is an ideal candidate for adoption. But suppose that, in the same hospital in which the infant is born, there are three other children, all five years old, who will soon die if they do not receive organ transplants. The newly orphaned infant turns out to have exactly the right tissue type: if it were killed, its organs could be used to save the three ailing children. According to the view I have developed, it ought to be permissible, if other things are equal, to sacrifice the newborn orphaned infant in order to save the other three children. [...] Most people find this implication intolerable, and I confess that I cannot embrace it without significant misgivings and considerable unease.
>
> (McMahon 2002 pp. 359–60)

Stephen Mulhall reviewed the book (2002) and opened his review with a discussion of the same passage. What is remarkable, Mulhall points out, is McMahon's candid admission that (i) his position does in fact have these implications and (ii) that the implications give him misgivings and unease. Such implications are a familiar problem for utilitarians, who usually invest great energy and ingenuity to deal with them and who never admit to being uneasy. One such utilitarian strategy might typically run as follows: 'If people are generally aware of the possibility of a newborn orphan being sacrificed, this will distress them and thereby bring down the aggregate utility. Therefore we can conclude that the sacrifice would not be the best option.' And this is often considered a fairly limp response because it does not eliminate the possibility of the orphan being secretly dismembered and more generally because it does not seem to take the *orphan* into account at all; its life depends solely on a sufficient quantity of distress among others. But McMahon does not provide this sort of qualification, either here or later in his lengthy book. The only way he can deal with his misgivings is to ignore them and to embrace the implications of his position with what he thinks is courage.

Now McMahon's approach would not be objectionable at all in most other academic disciplines. Imagine I tell a physicist that the table in front of me seems pretty solid, and he replies that in reality, 99 per cent of the table is empty space. This news might give me misgivings. And yet the physicist can legitimately ignore my misgivings because they are understandably based on an appearance of solidity. The physicist, with his expert knowledge about the nature of matter, can claim a certain authority to pronounce on the reality of the table.

With the ethical problem of whether to sacrifice the orphan, I claim that the typically scientific structure of enquiry and disagreement about the nature of reality and appearance is no longer appropriate. For McMahon's misgivings and unease *reveal the ethical reality* of the situation directly, a reality that is equally accessible to all of us and not just to putative experts. While he is reliably sensitive to that reality and admirably candid about it, he overrides the sensitivity because of a misplaced faith in the power of his theoretical models. And insofar as the models generate conclusions that go so egregiously against his misgivings and unease, they must be flawed at a very basic level.

This is a strong claim, and perhaps comes too close to intuitionism for comfort. Intuitionism is a generally discredited theory, involving (at least in its cruder versions) a 'sixth sense' by which we apprehend moral facts about, say, what ought to be done in a problematic situation.

The two obvious problems with such a theory are the widespread disagreements among people when making moral judgements, and the impotence of reason as a tool for one person to justify their own judgement or criticise another's. Before I offer further support for my account, however, I want to look at McMahon's larger project as a moral philosopher, rather than at his specific arguments. For if we can understand better what he was trying to do, we can evaluate the success of this book in those terms and ask more generally what philosophers should do when they encounter significant misgivings and considerable unease.

McMahon might defend his decision to ignore his misgivings on the following grounds: 'In any time there have to be people who have the courage to think the unthinkable, to follow logic and reason even to uncomfortable conclusions; for that is the essence of intellectual and ethical progress and the gradual overcoming of unreasoned prejudice, taboo and fear. Yes, I too have misgivings, for I too am susceptible to psychological pressures stemming from social conventions and from my upbringing, and I will gladly admit to them. Insofar as my misgivings can be supported by reason, I will endorse and follow them; insofar as they cannot, I will ignore them as best I can.' This line of defence dates back at least to Kant, who rejected moral sentiments as a ground for ethics precisely because they were capricious, conflicting and inconsistent.

Let us consider an example where we now believe that conservative misgivings were rightly ignored in the face of arguments for change and moral progress: the emancipation of the slaves in Britain in the early 1800s. For most of human history slavery was rarely challenged philosophically, and yet at this time a massive and seemingly irreversible transformation gradually took place. Although slavery continues to exist today, no serious Western politician would dare espouse it. The example of slavery is regularly invoked by philosophers not only as a blunt rebuke to the sceptical or subjectivist inclinations of first-year undergraduates, but also as a paradigm for animal rights campaigners, who look forward to a future when animals will be released from their present slavery. The details of the transformation are mostly a matter for historians, but McMahon could convincingly argue that detailed philosophical arguments along the way, formulated by people with the courage to think the unthinkable, helped shift the opinions of politicians and elites. The suggestion that a mere black man could, one day, be allowed all the rights of a white land-owner probably generated the same sort of misgivings and unease, even among the educated, as McMahon's proposed sacrifice does to the readers of his book.

There is a crucial problem with this analogy, however. The emancipationists expected their proposals to generate misgivings among their audiences, but they themselves were fully convinced about the rightness of their cause; they were convinced both intellectually *and* emotionally – and, retrospectively perhaps, we applaud them for that. McMahon has convinced himself intellectually, but his misgivings and unease remain. Does that not undermine something of his argument?

Not yet. McMahon could compare himself to the white parents in the 1968 film *Guess Who's Coming to Dinner?* The story is simple: although slavery is a thing of the past, racism is still rife in American society. The white parents in the film pride themselves on their liberality and have absolutely no problem with blacks among their friends and business colleagues. But when their daughter brings home a black fiancé, they are visibly taken aback. The film ends – as of course it has to end – with the gradual acceptance of the black son-in-law, and the underlying message is that stubborn, lingering prejudice can be overcome by reason and time and a bit of good humour. So maybe in the same way, McMahon's misgivings will ebb with time under the relentless flow of impeccable argument, and after we have 'tried' dismembering a few orphans we shall get used to it.

However, even if we describe McMahon's project as overcoming an understandable psychological reluctance, the analogy with emancipationism breaks down at two crucial points. First, the people *harmed* by emancipation were those who had invested in or relied on free slave labour and whose occasional and partial compensation would never cover the bleak financial future that awaited some of them. Some slaves were harmed in the short run by losing the reliable provision of food and lodgings, but in the longer term there are few ex-slaves who would claim this was not a price worth paying. In comparison, no amount of financial harm can equate with the harm of being killed, which would be the certain fate of all the orphan victims of McMahon's utilitarian calculations.

Second, it is hard to imagine McMahon starting a public *campaign* in the same way that the emancipationists did; hard to imagine him with a placard on the steps of the courthouse, demanding the sacrifice of the orphan – precisely because it starts to resemble nothing short of a lynch mob. Maybe he wouldn't demand the sacrifice as such, he would demand that 'everything be done' to save the lives of the three five-year-olds, that it was time for fearless thinking to overcome prejudice when considering the options available and making the best of an admittedly bad situation. But it is hard to imagine the campaign becoming widespread, precisely because of the misgivings and the unease to which he himself is prone. Indeed, while one can imagine an ideal world without any sort of racism, it is much harder to imagine a world where orphaned newborns are routinely sacrificed. Importantly, as Mulhall (2002) stresses, McMahon does not mention any parents. Even if the orphan's parents are dead, would their last wishes not have counted for anything? If not, that would undermine much of the solemn authority of wills. Nor does McMahon consider the parents of the three potential organ recipients. Might not they experience misgivings and unease upon hearing of the origin of the donated organs? McMahon presumes that they will be singularly delighted to have their child alive and well *at any cost*, but is that assumption valid? Parents will understandably place the welfare of their own child ahead of that of other children, but they may be less inclined to do so if the lives of the others are at stake.

(It might be objected that under certain extreme circumstances where survival is at stake, most individuals would kill rather than be killed, just as they would kill for their children rather than have them killed. This might well be true in a concentration camp, but it is important for the purposes of my book that such circumstances be seen as a rare exception. The shared ethical intuitions and concepts that I am interested in are paradigmatically acquired in more peaceful societies and are difficult to apply in confidence to more extreme situations.)

One can feel some pity for the destitute former slave-owners and for the disoriented ex-slaves; but this is nothing compared to the misgivings we feel for the victims of McMahon's calculations. Indeed, his use of the word 'misgiving' is disingenuous: the correct word to describe the appropriate emotional response to his brazen suggestion would surely be 'outrage'. That some slave-owners will face financial ruin may well be regrettable, but that a healthy orphan will be deliberately killed is a monstrosity. The pity that the emancipationists can feel for the ruined slave-owner will be limited by their knowledge of the moral status that the slaves had *all along*; the slave-owners were simply, perhaps innocently, blind to the implications of what they were taking for granted or even to what was just before them.

At the heart of the failure of the analogy between emancipationism and McMahon's proposal is an important point about how people change their minds on ethical matters – and this will be another theme of the book. I suggest that the British political class were not *persuaded* by the philosophical arguments adduced by the emancipationists, in the way that they might have been persuaded, say, about the wisdom of pulling out of the American War of Independence. Instead, they were *brought to see* black people as human beings, that is, as full members of the class of beings who were already considered as more or less equal (remember that not all white adults could own property or vote). This bringing-to-see is a much more complicated process, and I will examine it more closely in the next chapter.

Standing behind one's words

If McMahon's misgivings are strong enough to undermine his theoretical conclusions in the above way, then we must conclude that his proposal to sacrifice the orphan is not *serious*. What does that mean in the context of a discussion about ethics? McMahon's book is an excellent example of rigorous and detailed analysis of other people's arguments, and so in that respect is a serious work of philosophy, and I dare say widely admired as such. And yet I am suggesting he is not serious – indeed, cannot be serious – with regard to this particular implication of his position.

This goes beyond my earlier point about the public campaign. The key, once again, is to ask whether McMahon can really take his own conclusions seriously in his own life, outside the philosophy seminar. I don't mean take them intellectually seriously, I mean whether he can take his conclusions seriously in the real world, the world that the rest of us inhabit; whether he would *really* advise his obstetrician friend to go ahead with the sacrifice, whether he would really tell his wife that it was a good idea, and whether he would really go on national television to advise a change in health policy to solve the crisis surrounding the lack of transplant organs. And when I say 'really', I do not mean that he would say provocatively 'isn't it interesting to think *X*?' or 'imagine if we were to do *Y*?' By 'really' I mean that he would say: 'Here is my conclusion, and I genuinely think it is for the best. I stand behind my words. I fully endorse this advice.'

The phrase 'standing behind one's words' is Raimond Gaita's (2004, *passim*). Here is now a controversial statement: insofar as we suspect that he is not standing behind his words, then he is indulging in a mere intellectual game; if he is playing a game with such a serious subject matter, then he is not a serious *moral* philosopher. He might be very clever and very well-read; in any other branch of philosophy I'm sure he could generate elegant theories and powerful arguments. But moral philosophy is unlike all other areas of philosophy precisely because it cannot be isolated from the real world, it cannot be isolated from real people and it cannot be isolated from the moral philosopher's own life outside the seminar room. And if the moral philosopher is playing a game, it is not clear how reading him will help one to acquire a deeper moral understanding.

(Sometimes moral philosophy seminars turn into hackneyed debates between the Kantian and the Utilitarian, or between Absolutist and the Relativist. Much of the content of these debates is not serious, because there is no such thing as a complete Relativist or a pure Utilitarian, so why should we be interested in what they would say? Or rather, there may be a few examples of both, but they are surely very rare, close to madness and hardly pose a *threat* to ordinary morality. Although the debates might be a useful way to structure an admittedly complicated discussion, they can all too easily descend to the level of theatrical posturing and point-scoring.)

To some my critique of McMahon must seem either false or unfair, verging on *ad hominem*. Surely the arguments can speak for themselves; it should not matter who is writing them. McMahon may be wrong about many things, but his arguments deserve to be heard and engaged with. If there are errors of fact or inference, these can be pointed out, and McMahon will be only too grateful for the opportunity to improve the book. But how can it possibly be relevant to bring McMahon's private life into the debate about these arguments? After all, it is not hypocritical for a philosopher of science to question the reality of time and yet worry about being late for his seminar; part of his investigation into the reality of time will have to include an 'error theory' to explain why time does in fact seem so real to us. Such a theory will probably contain the thought that we cannot get out of time in terms of our experience or that our experience is essentially temporal (this goes back to Kant), but that we can still *think* about a world where time is an illusion. In relation to ethics, John Mackie (1977) famously advanced an error theory, designed to explain why, if 'there are no objective values' (the infamous words with which he opened his book), we persist in believing in them.

Insofar as McMahon and Mackie have recognised the 'error', their tone becomes one of claiming *expert knowledge,* and of enlightening the masses about such gross fraud, rather like Richard Dawkins's tireless efforts to expose religion as mistaken. Ordinary ethical unease and misgivings can be boldly ignored, because it is possible to discover how ethics 'really' works, in exactly the same way as an expert on teeth or cars or the nature of time can advise the layman what is really going on and what ought to be done. And yet is the same kind of expertise possible in ethics? I suggest not.[3] Consider the obstetrician, without any formal training in moral philosophy, who has no hesitation in criticising McMahon's proposed sacrifice of the orphan in the strongest terms. At the same time, it would be laughably inappropriate for McMahon, as far as I know lacking any medical training, to criticise the obstetrician's work to any detail. (There will of course be certain limits beyond which even a medically untrained person could legitimately criticise the obstetrician, such as when he uses rusty instruments or drops the baby on the floor.)

It is an essential feature of ethics that no one is immune to ethical criticism, neither McMahon with several weighty tomes in moral philosophy to his name, nor the obstetrician with all his medical knowledge. Clinical expertise, a set of relevant skills and knowledge, is something acquired in a specialised training, with a formal accreditation procedure. Without such training, a non-expert cannot rationally disagree with an expert (within certain limits): if a dentist tells me I have to have my molar out or if a car mechanic tells me I have to have my carburettor out, then I have no rational grounds by which to doubt their judgement. Any of my hesitations could in principle be met by the dentist's or the mechanic's exasperated statement: '*if* you had the relevant training and experience, I would be able to explain to you exactly what I am proposing to do, and why; and you would come to agree with me that it is for the best. In the meantime you will just have to trust me'. There is no comparable training in ethics and no criteria by which to distinguish expert from lay. As Raimond Gaita puts it (2004 p. 112), it is no accident that there are child geniuses in chess and violin-playing, but that there are none in ethics. It is no accident that there are no prizes or awards or quiz shows in ethics. It is true that there is such a thing as ethical education, but insofar as this is delivered systematically as part of a curriculum, it is delivered to all children; such that the adults that they develop into, insofar as they are generally mentally competent, will be *expected* to be fully conversant with the relevant discourse, to know how to take and give ethical criticism in the normal way: here there can be no distinction between expert and lay, but only between competent and incompetent. Every adult who commits adultery will be vulnerable to legitimate ethical criticism from every other adult. He may defend his adultery, but in doing so he is not invoking any expertise or specialist knowledge that he might have acquired academically or professionally.

Some philosophers resist this truth about the lack of ethical expertise. Here is David Brink:

> Lay persons are typically willing to defer to theorists or theoretical debate on matters scientific, but they seem largely uninterested in profiting from theoretical work that has been done in moral and political theory. [...] Most laypersons, even those with strong moral sensibilities, seem largely unaware of, or uninterested in, even the outlines of theoretical work in ethics.
>
> (Brink 1989 p. 207)

Brink seems genuinely surprised that people do not come to consult him about ethical problems. I find this quotation comical in its rather touching optimism, although it also suggests a simmering resentment that ethicists like himself have not yet been awarded the recognition and deference (and salary?) he clearly feels they deserve.

What McMahon is proposing is 'plain murder' – to use Anne Maclean's apt phrase when discussing a similar proposal by John Harris (Maclean 1993 p. 6). It is not that the option is unjustifiable, she continues; the question of justification does not enter the picture at all (p. 135). And this means that McMahon is *equally* vulnerable to ethical criticism in the philosophy seminar and in his own life; and that it is therefore legitimate and revealing to imagine how McMahon would fare if he advanced his philosophical proposal to a medical audience, to a prime-time TV audience, or to his own family.

The stuff of ethics is ordinary difficult situations, encountered by all of us in ordinary relationships with others. It is no accident that Socrates sees the fundamental question for ethics to be 'how should one live?' For him it was of course an immensely personal question and he excited and irritated his audience in large part precisely because he had the integrity to act according to his convictions and the gall to demand that others do the same. An economist can leave his desk and go home and become a friend or a father; a meteorologist can be brilliant at his job but hopeless at everything else in his life. There is plenty of room for expertise and theory in economics and meteorology, since that expertise and theory is essentially isolatable from the rest of life. One's ethical character, on the other hand, is something that one can never get out of, for it underlies all of one's interactions with others in all contexts. As such the ordinary ethical concepts that I use to make sense of my marriage can and should be the same ones that I use to make sense as best I can of the ethical dilemmas typical of modern medicine.

It is important to note what I am *not* saying: I am not saying that a generous husband will be a generous doctor. But the concept of generosity can be applied to both contexts in the same way. If my hospital colleague accuses me of lacking generosity, I will understand what he means, although I may of course deny the accusation when he spells out the reasons for making it. But if I care about his good opinion, I cannot just shrug it off, I owe him an explanation. When my wife makes the same accusation it might be in a completely different context, and for different reasons, but it is the *same* concept. Indeed, it is the same concept that I first learnt as a child.

This brings me to a justly famous quotation by Elizabeth Anscombe:

> But if someone really thinks, in advance, that it is open to question whether such an action as procuring the judicial execution of the innocent should be quite excluded from consideration – I do not want to argue with him: he shows a corrupt mind.
>
> (Anscombe 1958 p. 19)

The orphan in McMahon's example could be considered as up for judicial execution, although not for the usual political reasons but for apparently bona fide ethical reasons. McMahon, the fearless thinker, would not want to exclude *anything* from consideration: 'After all, what's the harm in considering? If it's not a good idea, I'll reject it in due course for the right reasons. But to exclude something from consideration smacks of dogmatism and cowardice.'

In response, Anscombe would develop this quotation to make the following two explicit points. First of all, she is not advancing an *argument*, such as an argument against capital punishment; she is stressing the wider meaning of the concept of innocence. By 'wider' meaning, I again mean one that includes what I have been calling the resonances associated with a successful ascription of the concept. McMahon would certainly accept that the orphan in his example is indeed innocent: both in the sense that he poses no direct threat to anyone and has done nothing to deserve death. But he would leave it there; nothing of ethical significance necessarily follows from ascribing that concept. Anscombe would claim that McMahon therefore does not really understand what 'innocence' means in its wider sense. If McMahon truly accepted that the orphan was innocent, this would restrict what could be done to it, such as deliberately killing it: such a restriction is not only ethical, it is also conceptual, built into the full meaning of the relevant concept. Traditionally, the *only* good reason for killing another human being is precisely when it is not innocent in the sense of posing a direct, and lethal threat to one (or posing an indirect threat through membership in an invading army).

Anscombe's second point is more important, however. The concept of innocence also restricts what can even be *contemplated* as an option. The mere contemplation of the possibility of deliberate killing of an acknowledged innocent already shows that one has not accepted the full import of the concept of innocence. It shows a corrupt mind in the sense that one is using the word 'innocence' irresponsibly, without a proper regard for the consequences of one's thoughts. What McMahon would call a 'mere' thought, to be rejected if it does not stand up to philosophical scrutiny, already reveals that he has crossed a line in his attitude to other human beings (again, insofar as he is serious). If the judicial execution of the innocent, or the instrumental pillaging of this orphan's organs, is a possibility in advance, even under very rare circumstances, and even to generate plenty of good, then there is no longer any such thing as an absolute prohibition against killing innocents. It is true that some philosophers, lacking any interest in the study of moral psychology or moral character, may not be troubled by such a conclusion. Anscombe's point is not one about a slippery slope, that a thought would increase the likelihood of action; rather, the mere thought is not as harmless as McMahon believes since it helps to corrupt the sensitivities to the resonances of the ethical concepts invoked in one's own life.

Conceptual revision

McMahon does discuss innocence at great length in his book, but always in the abstract. For example, on p. 401 he draws a distinction between an Innocent Threat and a Culpable Threat: 'Innocent Threats threaten without justification, but are nevertheless morally innocent because they have an excuse.' A plausible example of this might be a dangerous animal, which one would be justified in killing if it threatened people. But McMahon wants to apply it to threatening humans as well, as part of a search for a general justification to kill – and one of his conclusions is that a foetus can innocently threaten a woman's welfare. Although McMahon does not discuss it directly, one can presume from his remarks about foetuses that the new-born orphan would also be an Innocent Threat to the three five-year-olds, precisely because its continued existence puts their lives at risk through denying them the needed organs. But here he is taking the concept of a threat too far away from normal usage. After all, we can imagine a situation where my spending

£5000 on a new car could be causally linked to the death of a person in a third-world country whom the money, via a charitable intermediary, could have saved. Suddenly I am a threat to an awful lot of people.

The capital letters of 'Innocent' are revealing. McMahon would argue that he has to distil the complex real world into abstract types to discover the essence of the ethical reality in the situation. What he does not realise is that in order to revise the concept of 'innocence', he has to invoke other concepts such as 'threat', 'justification' and 'excuse', and take them in their ordinary senses in order to engage the reader. But the problem is that 'innocence' is already conceptually linked to these other concepts, so you cannot revise one without thereby shifting the whole web of meanings. And suddenly you have moved further than you intended. The worry here is nicely summed up by Stephen Mulhall (2002), and this reiterates a point I made earlier:

> Thought-experiments in ethics presuppose that we can get clearer about what we think on a single, specific moral issue by abstracting it from the complex web of interrelated matters of fact and valuation within which we usually encounter and respond to it. But what if the issue means what it does to us, has the moral significance it has for us, precisely because of its place in that complex web? If so, to abstract it from the context is to ask us to think about something else altogether – something other than the issue that interested us in the first place; it is, in effect, to change the subject.

The more abstract McMahon becomes, the less recognisable are the people he describes and the less he can rely on the readers according the same intuitive agreement. Suddenly threats become Culpable or Innocent, with nothing in between, a vision of the world as bleak and perhaps as dangerous as George Bush's.

Now it might seem that I am gearing up for the usual deontological attack on utilitarians, but I am not, at least not primarily. Certainly the deontological attack relies on similar intuitions in the reader that I am relying on. But my main question for McMahon is about why he doesn't trust the ordinary meaning of innocence, if it is not because it interferes with the philosophical conclusions he is already aiming towards. Consider the expressions 'innocent civilians' or 'innocent bystander', for example; here there is a very real implication of *inviolability*. When civilians are killed in warfare, it is always unjust. Despite the glib pronouncements about the inevitability of 'collateral damage' in a 'just war', there should never be a sense that it is fully justified to kill civilians simply because they live next to a military target, just as there should never be a sense that warfare is anything but obscene. Certainly the deaths are not justified *to the dead civilians*. Analogously, once the orphan is deemed violable, then *of course* the conclusion will be begged that its life will be outweighed by the three potential organ recipients. But this is hardly a philosophical breakthrough.

McMahon and others would complain that the persistence of ethical disagreement and dilemma would indicate that our existing concepts are too many or too vague or too inconsistent, and this calls for philosophical action to tidy things up. But it is also possible to reach the opposite conclusion: that the persistence of disagreement and dilemma indicates that our relationships are irreducibly complex, and that our first task is to do justice to this complexity while striving for a greater lucidity of detail. Despite the occasional confusion and uncertainty, however, the concepts are generally enough for ordinary people to make sense of our world and our actions, both now and in our remembered past, to communicate with one another about the world and about our actions, to praise, insult, worship, and demean; in short, to *lead a life* (this notion will be examined in greater detail in Chapter 6).

Bernard Williams (1985 ch. 9) famously declared that we have as many ethical concepts as we need. He was criticising attempts to reduce all the heterogeneity of our concepts to certain contrived terms such as 'right' or 'best' or 'harm' or 'rational'. Whatever the advantages of a common currency in which to measure and compare 'harm', one can surely be harmed (no inverted commas) in many ways. Physical injury is radically unlike remorse, for example; a hangover is radically unlike outrage, and yet all are somehow negative experiences. The concept of harm is taken to be essentially quantifiable. A little of it is bad, a lot of it is very bad: that's what simplifies the weighing. But consider the following situation: I slander a colleague unfairly in order to hinder his chances at promotion and thereby to improve mine. Later, I come to feel remorse for what I have done. Whether or not this remorse is pleasant or unpleasant is irrelevant; what is relevant is whether it is appropriate to the situation. If anything, remorse is a *good* thing to feel in response to my slander, because it shows that I have recognised the wrong I have done and that I have not become entirely insensible to the claims of decency and loyalty. In other words, it is a good thing, in this instance, that I suffer. But neither the suffering, nor the goodness of the suffering, can fit into the crude utilitarian categories of 'harm' and 'welfare' in any plausibly useful way.

Consider again McMahon's orphan; I suggested that the normal way for a utilitarian to defend himself against such counter examples would be to say that the harm caused to others (i.e. via their outrage at the sacrifice of an innocent life) could be weighed into the calculations and would probably tip the balance away from the sacrifice. I said above that this failed to take sufficient account of the *orphan's* point of view beyond his temporary status as the locus of do-gooding; but a more important problem is that it fails to take sufficient account of the outrage *as outrage*, rather than a generalised unpleasant emotion. Outrage is a response to an ethical wrong; if I ask the outraged person whether their outrage feels pleasant or unpleasant, they will probably not understand the question (assuming it is not a case of self-indulgent outrage by a person seeking the moral high ground).

So we have to look at the wider ordinary meanings of the central concepts of medical ethics. To close, let me look briefly at another key concept. The debate surrounding abortion is sometimes couched in terms of whether the foetus is a person, and therefore whether or not it is above or below the 'threshold of respect', as McMahon puts it. But the *ordinary* word 'person' is not something we would normally use with either a foetus or a baby. A person usually implies a context of adulthood, as in 'I met this strange person the other day, said he was an artist'. Or we can ask whether Mrs McGillicuddy is a sincere or a cowardly or a mischievous person. Or we can say 'what sort of person would go and slash the painting?' But an infant is a 'baby', a 'kid', or a 'brat' – not really a person. The concept of a person has become central to medical ethics, and especially to the debates surrounding the foetus, the severely handicapped neonate, the demented and the comatose. Some of these types of patients I will be considering in due course. The usual point of the revised concept of 'person' within the debates is to distinguish persons from non-persons, such that certain things – especially killing, or at least letting die – can thereby become ethically justifiable by comparing them to the killing of uncontroversial cases of non-persons (mosquitoes, say). Now, since personhood is not an empirical property, the next step goes, it would be good to find an empirical property on which it was necessarily and sufficiently based. In other words, we need to find some empirical property in virtue of which a competent adult human being is not a mosquito. We can then look for this empirical property in other beings – such as foetuses or severely demented patients – in order to resolve their ambiguous moral status. As such, the use of this concept works in a similar way to the ethical theory; we 'test' the concept on intuitively straightforward cases and then deploy it to resolve the cases where our intuitions are confused or contradictory.

When asking whether a comatose or severely demented patient is a person, we can look for empirical properties such as self-consciousness and rationality. These can then be expanded into properties such as the ability to have higher order desires (e.g. a desire to not have a desire to smoke), the ability to have complex language-based social relationships, the ability to plan for the future and regret the past, the ability to weigh options and make a decision about which would be the best. The next three-step is the crucial chunk of the argument: the comatose patient does not have empirical property *X*, therefore it cannot be a 'person', and therefore it is permissible to kill him, or rather, *it* (or, in a weaker version, it as at least permissible to accord its continued living less weight compared to other patients when allocating scarce resources or resolving conflicts). There is a different batch of empirical properties sought when trying to determine the moral status of the foetus at a given point in gestation. Here the paradigm is not a competent adult human being, but a healthy infant. So a foetus may lack a heartbeat, it may not have started kicking, organogenesis may not have completed, it may not be able to experience pain, etc., and *therefore* it is not a person and can be killed or at least accorded less protection when weighed against the mother's physical and psychological discomfort associated with continuing the pregnancy.

All this should be fairly familiar. The main point here is how utterly arbitrary the content of the concept of 'person' has become under revision, and how arbitrary the associated thresholds are. As soon as one departs from the *ordinary* meaning of 'person', then the game is wide open, and I can reach any philosophical conclusion by stipulative definition. After all, I can take *any* properties that healthy infants have but which foetuses lack – weighing more than 2.5 kg, having eyelashes, being able to see etc. – and call *these* essential to personhood and design my abortion policy accordingly. But what on earth does this achieve?

2
Ways of Seeing

In the last chapter, I introduced the important place of the spontaneous intuitions and conceptual resonances of ordinary people when facing ordinary situations, as contrasted with the theoretical conclusions of moral philosophers. I admitted that my approach seemed too intuitionist, and this led to the problem of accounting for and resolving ethical disagreement. In the first section of this chapter, therefore, I would like to examine the phenomenon of ethical disagreement directly. As with the first chapter, this chapter will continue to lay important groundwork for the more applied discussions in Parts II and III, and for that reason will seem to move some distance away from the more familiar topics of medical ethics.

What does an ethical disagreement typically look like, and what does the (satisfactory or unsatisfactory) resolution of such a disagreement look like? Take two individuals, *A* and *B*, who have a *prima facie* ethical disagreement: *A* claims *p*, and *B* claims *~p*, where *p* characteristically involves some proposition that describes an action as being ethically best or right in a given situation, or describes a person as ethically commendable in some way, or describes a fact as ethically relevant in some way. What might be going on here?

1. It might not be a *disagreement* at all, but an exchange of aesthetic or gastronomic preferences. If you prefer wine and I prefer beer, we are not disagreeing about anything.
2. It might not be an *ethical* disagreement at all, but a trial of power. The words just disguise the intentions of the speakers, which is to make the other do something that he would not normally be inclined to do.
3. It might be an ethical disagreement, but one that could in principle be satisfactorily *resolved* by the patient exchange of relevant non-ethical

facts or by the correction of inferential errors or by some political settlement involving compromise or compensation or force.

Optimists will see this as the end of the list. However, I will not be considering any of these possibilities further here. Instead, I am interested in genuine ethical disagreements between two sufficiently well-informed parties of good will. By 'sufficient' here, I am assuming the basic plausibility of often reaching a point where no further non-ethical information is relevant, and the disagreement 'jams': the fundamental disagreements between the *Daily Mail* and *Guardian* readers (i.e. between right and left of the political spectrum) are the most obvious examples, as well as some of the core disputes that define the subject of medical ethics. There are three types of such jammed disagreements.

4. *A (contingently) resolvable disagreement.* One person comes to agree with the other, but not through the process described in 3. Instead, one disputant comes to see – or is brought to see – the situation through the other's perspective, perhaps through the deployment of helpful analogies and comparisons, of illustrative examples etc. Whether and to what degree such a change involves rhetoric of the pernicious kind will be examined in a moment.
5. *A (contingently) irresolvable disagreement.* The efforts described in 1–3, as well as in 4, fail. However, each can acknowledge the other's opinion as intelligible and even respectable, although they will 'agree to disagree'. Sometime the disagreement will be harmlessly mysterious; at other times it will require efforts to explain the contrary opinion away with reference to some character-based or role-based determinism: 'You would think that, given that you're a bank manager!'
6. *A radical disagreement.* Not only do the two disputants disagree, but the position of each is not fully intelligible to the other as a coherent ethical position. Moreover, one disputant finds he is unable to explain away the other's position with reference to the other's character or role. As a result it becomes difficult to 'share the same ethical universe' with the other, to use Cavell's expression. (One example of this might be a doctor's conscientious objection to performing abortions, which I discuss below.)

It is important that I have used the word 'contingently' in 4 and 5. The point is that a single disagreement may *turn out* to be resolvable or irresolvable, and there is no way to predict this in advance – we can only wait and see. This is not some banal reminder of the psychological limitations of the disputants, but an important characteristic of these ethical disagreements. Mainstream philosophy models ethical disagreements on disagreements about matters of fact. When I claim that there are two chairs in the next room and you claim there are three, at least one of us must be wrong – but we must *both* accept that the disagreement can be resolved by going next door and having a look. Once there, there is 'nothing left to think' (to use David Wiggins's phrase) but that there are the number of chairs we see. We are answerable to the facts, and there is no room for contingency about what to believe, unless we allow one of the disputants to have some perceptual or cognitive impairment, or unless we have a new disagreement over whether this bench is or is not a chair. In the same way then, the mainstream moral philosopher will, optimistically, claim that all ethical disagreements are type-3 and essentially resolvable, regardless of the contingent identity of the two disputants, because the disagreement is answerable to a singular truth of the matter. I do not dispute the existence of type-3 disagreements, but claim that some disagreements are type-4 and type-5, and that these reveal more about the nature of ethics. And it is these disagreements that I want to examine in greater depth.

The vegetarian and the carnivore

I want to consider an ethical disagreement between *V*, a vegetarian, and *C*, a carnivore, over the permissibility of eating animals. I have deliberately chosen this debate rather than a debate from medical ethics because it is conceptually simpler and less is at stake, and so it will be easier to make my philosophical point about the nature of the disagreement. In the later chapters, I can then bring some of points to bear on more familiar debates such as those of surrounding abortion and euthanasia.

V and *C* agree in many of their ethical judgements, but clearly they have one source of persistent disagreement, and that is the moral status of animals. I want to ask what it would mean for one to *persuade* the other to change his mind on this issue. Here is the situation: (i) *V* claims that killing animals for food, when not a matter of human survival, is ethically impermissible; (ii) *C* claims that as long as the animals are well treated during their lives (e.g. on organic farms) and killed humanely, then it is ethically permissible to eat them, and therefore permissible to breed them to that end; (iii) the two of them are visiting a local village zoo with small domestic and local forest animals, including a common chicken; and (iv) I shall not consider the rejection of animal *products* (eggs, milk), nor shall I consider factory farming or blood sports or experimentation on animals here.

The disagreement is not one of preference – both parties are worried about doing the ethically right or at least permissible thing. What's more, V is determined to persuade his friend C to see the light, and C is willing to make the effort to understand V's position (he will not change his mind just to please his friend).[1]

V might start with some of the classic arguments in favour of restricting the way we treat animals: (i) that humans don't *need* to kill animals, since they can live perfectly well as vegetarians; (ii) that animals share certain capacities with humans, especially the capacity to suffer, and therefore they deserve to be treated in the way that we treat other humans; (iii) that mentally, animals are no different at least from infant or senile humans, and therefore deserve the same respect and protection; and (iv) that there are grave contradictions between our carnivorism, on the one hand, and our attitudes to domestic pets and to the talking animals in our children's fairy tales, on the other.[2] C can respond in one of two ways. Either he can advance his own ethical arguments and declare that he finds them more persuasive – he might say that animals and humans have their specific places on a natural (or divinely-sanctioned) food chain, that higher animals have no more moral status than lower animals or even man-made machines – or he might admit that V's arguments are very powerful, but he is not sufficiently persuaded to overcome his delight at eating meat. Either way, the disagreement has jammed, and no other facts, evidence or arguments can plausibly be adduced by either party.

So here's a bold suggestion to break the impasse: what if V takes C to a slaughterhouse? Perhaps C has not hitherto made the connection – *at a deeper than merely intellectual level* – between the harmless living creature and the tidily packaged lump of meat in the local butcher's. Or we might put it this way: C has been told, but not shown the truth. He knows that chicken meat comes from chickens, of course, but in his primary school he was never asked to draw the intermediate stages between the happy farmyard and the dinner table. So it could be said that he does not *know* where the meat comes from, with the implication that *if* he really knew – knew more than the mere proposition, knew enough and in the right way to be *moved* – the chicken would change its ethical shape, so to speak. Again, in one sense this is ridiculous, the chicken remains a chicken whatever happens to C; in another sense, however, the chicken would become a very different object within C's perspective, because it would now come with new prohibitions. This is not a matter of 'projecting' the prohibitions onto the chicken, for the concept of projection is only something that can be applied by a third-personal observer. If we want to make sense of the first-personal experience, then we must recognise that the concept of projection does not itself enter that experience: the prohibitions are out there, to be discovered.

We are back to this distinction between different kinds of knowledge again. Until the visit to the slaughterhouse, C's views on the moral status of animals were too 'cheap', since he could effectively ignore the origin of the steak, and the meat industry colluded with that ignorance by hiding the slaughterhouses and packaging the meat so tidily. A similar kind of knowledge-and-ignorance seems to exist with cigarettes, which everybody 'knows' are harmful, and yet few people know what it is like to develop smoking-related illnesses such as lung cancer. Would a tour of the lung cancer ward at the local hospital help smokers to quit? It might help some, certainly. But given the number of doctors who continue to smoke, it will never help everybody. A third example might be the cheap textiles and shoes that we have come to take for granted in the west, without really knowing the awfulness of the third-world sweatshops that produce them. Once we distinguish between these kinds of knowledge, between what might be called propositional and experiential knowledge, lots of examples start to present themselves. Suddenly we meat-eating, cigarette-smoking cheap-textile-wearers become irresponsibly cavalier about the prerequisites and consequences of our actions. Even if it is only I who will suffer from my smoking, my other actions will cause animals and sweatshop workers to suffer. On the other hand, there is something unseemly in wallowing in such a massive burden of guilt, and something akin to ethical bullying in imputing such guilt on others, since it would be hard to know where to stop in this massively interconnected world, without becoming a self-sufficient recluse.

Either way, the dominant fact about this kind of deeper knowledge is its inescapable contingency. Perhaps the slaughterhouse could join the experience of the living and the dead animal together, suitably accompanied by the blood and the cries and the twitching freshly killed corpses. This might 'work', C might be shown rather than just told, converted rather than persuaded; *then again, he might not*. And if it doesn't work, there is probably little more that V could do except say helplessly, 'but don't you see?' However, at least C will probably be able to say that he can better *imagine* what V sees; that he knows better 'where V stands' or 'where V's coming from'. What is important here is to realise that the failure to convert *C* does not necessarily reveal an irrational stubbornness or other culpable failure on *C*'s part.

Let us say that the slaughterhouse visit converts *C* to vegetarianism. Is this then a legitimate form of mind-changing? After all, *C*'s rational will seems to have been bypassed, and this would bring the process close to such illegitimate mind-changing processes as hypnotism and brainwashing, or at the very least spin and sentimental pleading. I would suggest, however, that philosophers have been naïve in assuming both that arguments can be presented with perfect neutrality and that the *only* alternative to perfect neutrality is brainwashing. In truth, many philosophers are justifiably proud of their elegant writing style and pithy examples, their careful emphases and juxtapositions, and there are different degrees of 'rhetorical flourish', some more legitimate than others.

Consider the use of examples in ethical arguments. It is tempting to assume that an example is meant to illustrate a principle or a theory in action, or clarify a relevant distinction, without contributing anything to the theory or principle, and without substantively advancing the argument. The whole tone should be parenthetical, as if allowing lesser minds to keep up with the pace. As Cora Diamond (in 'Anything but argument?', 1991) has pointed out, however, philosophers have typically been reluctant to see examples as a vehicle for *indirect* persuasion, or rather, indirect conversion, or moving. What I have in mind is how one can be moved by a work of literature or film, moved in a way that would involve adopting a new ethical perspective in a situation, with certain new ethical judgements. Think of a person 'converted' to medicine by watching the hospital television drama *ER*. Again, such a conversion invites suspicion insofar as *ER* simplistically glamorises hospital work, but nevertheless the convert need not be *embarrassed* by his conversion; whereas, he would be embarrassed by evidence that he had adopted a belief by hypnosis, and this embarrassment would undermine the belief. As such we can distinguish between philosophically legitimate and illegitimate forms of indirect conversion by what I will call 'transparency of method'. A social reformer should not be embarrassed to admit that he acquired many beliefs about justice while reading Dickens, and *C* should not be embarrassed to attribute his conversion to his visit to the slaughterhouse. Indeed, it goes further than this: without being moved by something like reading Dickens, the social activism risks smacking of self-righteous posturing. As such, appeals to the heart and the imagination

are just as legitimate, within limits, as appeals to the mind; and show-
ing can be as legitimate as telling.

But does such conversion not leave too much open to the heart and
the imagination? Could one not be converted to a more nefarious ethical position? After all, my choice of the disagreement between the vegetarian and the carnivore is hardly the most important of ethical issues,
and so many would think that it does not really matter whether C is
converted. However, imagine the example of a white doctor, with a
solid humanitarian impulse, going to work in a crowded and poor
inner city dominated by ethnic minorities. After a while, he is 'struck'
by what he sees as their laziness, their resentment and envy, their
sexism – and he is 'converted' to a racist position. This is a person who
would never have believed any *arguments* supporting such a position,
but who has been moved by bitter experience towards a new ethical
outlook.

Certainly such conversions do happen, but what is important is that
the doctor's conversion itself is not the end of the story. There is still
room for reflection on the experience, on the conversion, and on the
new inclination to racism. And it is possible to refrain from *endorsing* it,
because it fails to conform to an ethical ideal that he still cherishes. The
process of endorsement is important. The Carnivore can be brought to
see that meat eating is wrong, but rather than simply change his eating
habits, he will endorse the judgement by adopting (rather than absorbing)
the reasons for believing that meat eating is wrong. The doctor will resist
his new racism by refusing to offer reasons in support of it.

Similarly, one has to be careful not to take the conversion too far.
Animals are still *only* animals. The disagreement between V and C was
strictly about the permissibility of eating them, and there was an important agreement between both parties that humans no longer have to eat
animals for nourishment. Some animal rights campaigners err, however,
in imputing a status to animals that is sometimes too close to that of
humans: if faced with a choice between saving a senile old human and
an intelligent young dog from the proverbial burning building, they
might be tempted to give priority to the dog on account of its greater
cognitive capacity and better life prospects. But this would be monstrous, because *it's just a dog*. It would be equally inappropriate to organise a lavish funeral for a dog.

I will inevitably be accused of specieism – another loathsome
neologism – and of course I am. But the animal rights campaigner anti-
specieist goes too far in ignoring what status animals have had in our

lives *until now*, a status that cannot be overturned in a day. The animal is not a neutral being, whose moral status can be deduced from its empirical properties. As Cora Diamond puts it:

> A difference like that may indeed start out as a biological difference, but it becomes something for human thought through being taken up and made something of – by generations of human beings, in their practices, their art, their literature, their religion, their ethics. [...] It is absurd to think these are questions you should try to answer in some sort of totally general terms, quite independently of seeing what particular human sense people have actually made out of the differences or similarities you are concerned with. And this is not predictable. [...] We are never confronted merely with the existence of 'beings' with discoverable empirical similarities and differences, towards which we must act, with the aid of general principles about beings with such-and-such properties deserving so-and-so.
>
> ('Experimentation on animals', 1991 p. 351)

This is a very important quotation for my purposes, since it emphasises the role of inherited social meanings in ethics. When debating what ought to be done, we always have to start with what *is* done. When we later enquire about the status of the foetus, we have to start by asking what that status is in *normal* pregnancy, and what role normal pregnancy and birth play in ordinary lives. Again, I will always be at pains to stress that this does not collapse into cultural relativism or conservatism; instead, it is a thesis about the meaning of concepts. And the meaning of the concept 'animal' is shaped precisely by the practices of 'generations of human beings'. There is room for the meaning to change, but only slowly, as more and more people are contingently brought to see what they only knew before. This is the truth about the emancipation movement. And any discussion of abortion and euthanasia has to start with what generations of human beings have 'made' of pregnancy and death 'in their practices, their art, their literature, their religion, their ethics'. Only from such a starting point can we be confident of using a word like 'cadaver' with its full meaning. As such, the whole effort to ascribe *rights* to animals on the basis of their empirical properties is wrong-headed. And, I shall argue, the whole effort to ascribe rights to foetuses is similarly wrong-headed. Neither of my methodological claims undermines philosophically legitimate beliefs that meat eating or abortion is ethically wrong, however.

Let me finish off this section with a further example of a disagreement, this time between a 'doctor' and a hospital 'manager'. The doctor has an employment contract with the Hospital Trust, and the manager is partly in charge of making sure that the doctor complies with that contract. One day the doctor's infant son becomes ill, so the doctor decides to stay at home to look after him. Now the employment contract stated precisely that in exchange for such-and-such a salary and benefits, he was committed to perform such-and-such tasks; he could only miss work for 'good reasons'. Most of the time, the doctor would be in entire agreement with the manager about what would and would not constitute a good enough reason to miss work: a family bereavement would, and a favourite soap opera would not.

But within such limits, on this specific question of his son's illness, let's say they disagree strenuously. 'Of course I have to stay home with my son', says the doctor. 'Of course you have to come to work and see your patients', says the manager. And here's the point: even if they were as informed as one could expect them to be (of implications, consequences etc.), even if both parties were as open-minded and good-willed as one could expect them to be, even if the contract between them were as explicit and precise as possible, it is surely easy to imagine such a disagreement persisting to the point where even *hypothetical* or *ideal* convergence could not be rationally expected as necessary. Optimists and skilled professional negotiators might well believe that some sort of compromise is always possible with enough imagination, or some sort of compensation can be offered to the injured party. In one sense, of course, the disagreement would eventually have to be settled, but such a settlement would be *political*; and would fail to persuade the losing party – where by 'persuade', I mean in the full sense of inducing rational assent. One party would have to threaten or entice the other party, thereby reorienting the disagreement, and the outcome of such a conflict would then depend on the power relationships and bargaining positions between them. They may well continue to work together, each mumbling about the other's misguided priorities. This one conflict may reveal longer-term patterns, making the one a 'family man' and the other a 'career man'. But can either be said to be *right*?

This is not to deny that either the doctor or the manager might later come to change their minds, that they might be brought to see the situation from the other's perspective, but such an event could not plausibly be interpreted as convergence upon the singular truth (of whether the doctor was or was not justified in taking time off work) because it would lack the necessity generated by rational answerability to such truth. And if such a persistent disagreement is legitimately possible in the relatively explicit contractual relationship of the doctor and the manager, imagine how much more disagreement is legitimately possible in the ordinary non-contractual relationships we have with others, from complete strangers to intimates. While the doctor and the manager were forced to work together in the short term, however grudgingly, when there is no such force we often do not even try to resolve a legitimate ethical disagreement – we simply walk away. If it happens too often and too seriously with a specific individual, we start avoiding them, thus making the implicit judgement that we no longer want to share the same ethical universe with them, as Cavell puts it (1979).

When I am ethically perplexed, I can seek advice, and there may be plenty forthcoming. But I cannot abdicate the decision to someone else in a way that would shift responsibility and blame onto that person, in a way that I can onto the dentist or mechanic. After all, when I receive even the best advice I shall have a new problem about whether to accept that advice, and to make it *mine*, at which point it is no longer your advice but my decision. As Rush Rhees puts it, in an ethical problem nobody can tell me what I ought to do, not because the subject is difficult, nor because it depends on something to which I alone have access; 'mainly it is that the question is not answered until I answer it' (Rhees 1999 p. 75).

Conscientious objection

Sometimes the ethical disagreement between two people might take the form of a certain 'moral incapacity',[3] as when the Vegetarian declares that he 'cannot' eat meat. A famous example from the world of medicine is that of the doctor who declares a legal conscientious objection to performing abortions. I will leave a more detailed discussion of the abortion debate until Chapter 5, but I introduce this notion here because it will usefully bring together a lot of the themes of this book. For in citing a conscientious objection, he is declaring something essentially personal about himself, rather than making a public criticism of an NHS policy or of his colleagues, and rather than inviting a public philosophical debate. At the same time, as with all cases of genuine moral incapacity, this is not a *merely* personal issue for him, in the way his preference for beer over wine is. For his experience of the situation is that of an objective prohibition. Despite his objection, however, he *is* able to work for an NHS that routinely and publicly provides abortions, more or less on demand; indeed, he is able to work alongside colleagues who carry out the procedure.

Legally, conscientious objection to abortion is explicitly governed by the Abortion Act 1967, section 4,[4] where NHS medical staff are excused from 'participating' in the process of approving and performing abortions. There are two restrictions to this clause, however. First, the abortion must be performed by any suitably qualified medical practitioner (i.e. including the conscientious objector) if it is 'necessary to save the life or to prevent grave permanent injury to the physical or mental health of a pregnant woman'. Second, the objecting NHS medical practitioner must give the woman the name of a colleague whom he knows will be more likely to agree to the request. (The issue will be slightly different for doctors working in private medicine, and I shall not consider them here.)

There is already a good deal of philosophical literature on conscientious objection to military conscription. Interestingly, the Abortion Act was passed at the time that such debates were raging over American participation in the Vietnam war. However, objecting to abortion is different for an important reason, and that is that in 1967 there was growing opposition to the war, while in 2007 any resistance to the Abortion Act has dissolved in the 40 years of its validity. In hindsight, the Vietnam war has come to be a classic example of an unjust war, despite being legally approved by the American Congress; and conscientious objection seems like a coherent response as a result, helping to turn the tide. No such hindsight decisions have been seriously made about the Abortion Act in a way that would generate a real possibility of repealing it.

In a usefully controversial article, Julian Savulescu (2006) argues that the conscientious objection clause in the Abortion Act should be removed, for a number of reasons. The main argument is that the NHS is governed by law, and that the law should be applied consistently by all practitioners of the NHS. Conscientious objection throws an obstacle into the consistent allocation of treatment, and an obstacle to a patient who may be in a particularly frightened, vulnerable and confused position. At the extreme, a conscientious objection can be particularly damaging in a remote rural area, where the patient would then be forced to travel far in order to get the standard NHS treatment to which she is entitled. If the objecting doctor wishes to attempt to change the law, then he should do this in the usual and proper manner, for example by lobbying parliament. But so long as he works for the NHS, then he should abide by the wishes of the people as expressed through parliament – and he cannot deny that attitudes to abortion *have* shifted in the past 40 years. There is no obligation for him to work for the NHS; he could work for a private healthcare organisation and legitimately refuse all non-emergency requests for abortion.[5]

Second, in making an exception for himself, he is creating a dangerous precedent; what if another doctor conscientiously objected to treating a patient of the 'wrong' skin colour? After all, the law stipulates that 'the burden of proof of conscientious objection shall rest on the person claiming to rely on it', but provides no standards by which to assess such proof. Such standards would be notoriously tricky to draw up. The NHS has an explicit commitment to fighting all forms of racial discrimination, and a racist but otherwise skilled doctor can be legitimately expected to keep his opinions to himself or leave the service. Racism is perhaps a less plausible example in the sense that there is no longer any room for it in respectable public discourse, so Savulescu considers a better example: what if a doctor conscientiously objects to treating a patient of over 90 years of age, supporting his objection by the belief that the patient has had a 'good innings' and therefore does not deserve the scarce resources which the NHS has committed itself to providing to all patients on the basis of their health needs. Such a 'good innings' position is more philosophically respectable, but it is not yet NHS policy; and until it is, no doctor can legitimately refuse to treat on those grounds alone.

Third, there is a peculiar futility about conscientious objection, for the objector himself has to provide the name of a colleague who he knows will authorise the procedure. So the objector is implicated. And even if word gets around that the objector won't sign the form, and pregnant women stop asking him to do so, the objector still knows that any woman will be able to get an abortion anyway; so long as he refuses to lobby parliament, then his opinion won't change the system. And there is something odd about passing on the dirty work to his colleagues because he wants to keep his hands clean.

These three arguments, written by a qualified doctor (so perhaps with greater authority than a mere philosopher), seem to make a very strong case for eliminating the conscientious objection clause. At the very least they merit a sensible, measured discussion. What is striking, however, is the huge and angry response (mostly from doctors) to Savulescu's article among the readers of the *British Medical Journal*: 47 pages' worth, and all but two of the responses critical. In addition, one reader declared his intention to cancel his subscription to the journal, another that he had never read anything as awful in 50 years, a third that he was made physically sick; and many intimated that Savulescu would have supported the Nazi experiments on concentration camp inmates. This was truly a remarkable response, and clearly a nerve was struck. What are we to make of it?

There are three things about the objectors' position, insofar as we take it seriously, that are distinctive. First, it directly challenges the assumption, widespread in mainstream moral philosophy, that a single, ethically neutral description is available; instead we have another case of irreconcilable 'ways of seeing' and the limits of attempts to persuade. Second, it challenges the assumption, again widespread, that a single 'space' of reasons or arguments exists between all respectable disputants, such that *prima facie* inconsistent reasons and arguments can be weighed against each other in order to determine the overriding reason or argument that would then be binding on all concerned (or to determine what ought to be done 'all things considered'); instead, we have strong convictions supported by ways of seeing rather than by reasons. Third, we have people with strong convictions who refrain from condemning their colleagues and their employer for actions that they can only consider deeply immoral.

Imagine a doctor is asked to authorise an abortion (and to simplify things, let us assume the woman's life is not at risk, and there is no known foetal defect). He refuses, saying 'I'm sorry, I can't do it'. Now moral philosophy is normally thought of as a matter of deliberation and choice (on the basis of reasons), and yet here is the objector denying that he has any choice. Is this a cognitive failure of some sort – are there options that he is not aware of? Either way, how can it be ethically admirable, coming as close as it does to a statement like 'I'm sorry, I couldn't help it', the classic device to *evade* responsibility?

Let us be clear what the objector is *not* saying. He is not saying he is physically unable to sign the form, in the way that he might be unable to run a mile under four minutes. He is not saying that he is psychologically unable, in the way that he might be unable to bungee jump. And he is not saying that he is hypothetically unable, in the way that he might be unable to stay at work beyond five o'clock because he wants to prepare for the theatre. What is revealing about all three types of alleged non-moral incapacity is that certain responses then make sense. With the theatre date, we could ask if the theatre is *really* so important that he can't make the five o'clock meeting. With the four-minute mile and the bungee jump, we could say 'grit your teeth', or 'come on, everyone else has done it', or most importantly, '*try*'. It is as if some extra effort of the will is required, perhaps gradually over the longer term, to overcome an obstacle which does not seem to hinder others.

And if the declarer tries and succeeds, this will falsify his earlier denial: 'see, you can do it after all'. (Hence the expression: 'every man has his price'.)

However, when the conscientious objector declares that he cannot sign the form, it would not make sense to ask him if the matter was 'really' so important, nor would it make sense to urge him to 'try'. These attempted reinterpretations fail to appreciate the objector's perspective on the situation; that for the right sorts of reasons, the doctor will not even *try* to overcome his incapacity. Is the doctor's incapacity anything more than the Categorical Imperative or some similar sense of ethical obligation that binds me, as a rational agent, regardless of my inclinations? The answer is revealed by noticing that the doctor does not say 'abortion is just wrong, and it should not be done'. The essential feature of the Categorical Imperative is its universalisability: in discovering what I ought to do or not to do, I am implicitly discovering what ought to be done by all rational minds in this situation. The doctor, I suggest, is not interested in what other people can or ought to do; he really is declaring something only about himself, something deeply personal about his experience of the circumstances in which he finds himself, something that he might not even have anticipated until he got there. At the same time, it is not a *merely* personal whim or preference, as evidenced by the fact that he is willing to risk angering his patients and his colleagues. (In the context of objecting to military conscription, the objector even risks a gaol sentence.)

Now imagine that the objecting doctor is an obstetrician. He is well known for refusing to perform abortions, and he dutifully refers such requests to his colleagues. However, imagine that he is the only obstetrician on duty, and a woman is brought in with serious injuries from a road traffic accident. It turns out she is 18 weeks pregnant, and is losing a lot of blood because her injuries have caused the placenta to partly detach. At 18 weeks, the foetus is not viable, and in order to get access to and clamp the blood vessels of the uterus, as fast as possible, the obstetrician decides that the foetus has to be removed and killed.

When there is such a direct threat to the mother's life the doctor is legally required to abort the foetus; but let us say that in addition to that the doctor in this case agrees that the abortion is necessary. Now such a situation would not seem too problematic – in terms of ethical theory – to a mainstream philosopher like McMahon, so long as the threat to the woman was serious and credible, so long as the response was clinically appropriate, and so long as the obstetrician had no ulterior motive in killing the foetus, all of which I shall assume. The abortion, it would be said, was *justified*. But would the objecting obstetrician see it that way?

I suggest not. In discussion afterwards, the obstetrician will say that he 'had to' or 'was forced to' do it, but what he had to do was still, within his perspective on the world, an ethically wrong act. While necessary, it was unjustifiable. Indeed, not only did the obstetrician not consider the abortion right or justified, but he did not attempt to reorganise his principles or maxims to allow this sort of exemption. He did not say to himself: 'Henceforth I will be a conscientious objector unless I find myself in situation *X*'. After all, the whole point of the conscientious objection would be lost if it were thought of as subject to qualification in this way. And yet it is tempting to insist that obstetrician *must* have seen the action as right or best *in some way* (or: 'all things considered') or that he was somehow pursuing his conception of the good – for otherwise he surely wouldn't have done it. But this insistence ignores the objector's own view on the matter, and how troubled he is after the event. For while he never has any doubts that he did what had to be done; he finds that he cannot satisfactorily *justify* it, either to himself or to others. And yet he will be an even more determined conscientious objector than before.

Let me conclude this section with a striking literary example, featuring not a foetus but a fully grown woman who is the hero's daughter, in order to show how it might be necessary to kill a person who is deeply loved, and about whose moral status there can be no question. Agamemnon, in the version of the story by Aeschylus, has been becalmed with his army at sea on the way to attack Troy. The gods demand the sacrifice of his daughter, Iphegenia, before the wind resumes. Otherwise, Agamemnon, his daughter, and the entire army and crew will all perish of slow starvation. Once he learns of what is required, Agamemnon does not really stop to deliberate, nor does he really hesitate; instead, he pauses to appreciate the appalling crime he is about to commit, has to commit, a crime that he knows can never be justified, let alone expiated. Bernard Williams describes it movingly:

One peculiarity of [extreme cases of moral conflict] is that the notion of 'acting for the best' may very well lose its content. Agamemnon at Aulis may have said 'May it be well', but he is neither convinced nor convincing. The agonies that a man will experience after acting in full consciousness of such a situation are not to be traced to a persistent doubt that he may not have chosen the better thing; but, for instance, to a clear conviction that he has not done the better thing because there was no better thing to be done. [...] Rational men no doubt pointed out to Agamemnon his responsibilities as a commander, the many people involved, the considerations of honour, and so forth. If he accepted all this, and acted accordingly: it would seem a glib moralist who said, as some sort of criticism, that he must be irrational to lie awake at night, having killed his daughter. And he lies awake, not because of a doubt, but because of a certainty.

(Williams 1973 p. 173)

The limits to a philosopher's authority

In this final section I want to consider one further problem with the mainstream approach to medical ethics, and that is a problem of *authority*. In the previous chapter, I considered the difficulty of speaking of an ethical expertise analogous to expertise in medicine or in car mechanics. I now want to suggest that it is even more difficult to speak about expertise in *medical* ethics among philosophers who are not themselves healthcare professionals. Clearly there are many philosophers who think they do have the authority to offer something to doctors, and in the United States there are even hospital 'ethicists' with bleepers who can rush to the bedside to advise on the ethically most appropriate course of action.

Right away let me propose a general ban on the word 'ethicist', precisely because of its misleading phonetic associations with genuine experts such as anaesthetists and physicists, or with ideological proponents such as Communists or Baptists, and because of the whiff of preachy self-righteous moralism. Besides, it's an ugly and unnecessary word in an age of rampant neologisms: why not just 'philosopher'? (Although that too has unfortunate connotations, I admit.)

The more serious concern I have is that the mainstream philosopher does not sufficiently appreciate just how extraordinary the world of medicine is, and therefore just how extraordinary the issues are that take place within that world. Most of his experience will typically be from brief episodes as a patient, from horror stories related by a doctor friend, and from hospital television dramas. And yet despite this lack of experience he seems to show no hesitation, no humility, before plunging into the big questions of life and death. The risk is that the big questions will be more about *abstract* questions of life and death, rather than about the questions actually faced by healthcare professionals and by patients and relatives. I believe that the experience of working in a healthcare environment gives one a special kind of knowledge that cannot be gleaned from books, and that this knowledge is directly relevant to any deeper philosophical understanding of the complex issues in medical ethics. I want to introduce this notion here, and then examine it further in Chapter 9 by looking at the attitude to human bodies within the medical world.

In many ways, of course, medicine is just another university course, and just another career. The medical training occupies no special place in the university prospectus, and the hospital staff are organised along the same bureaucratic lines as any large institution. Off-duty, doctors and nurses dress and speak and drive like the rest of us. But this similarity occludes striking dissimilarities.

Consider the GP: only in one other context (sexual pursuit) would we undress in front of a complete stranger in order to allow him to touch us, sometimes intimately. Some of the information requested by the GP, we wouldn't give to our closest friends or family, let alone a stranger: problems with drink, sex and continence, for example. This represents a huge amount of trust, and gives the GP a huge amount of power, whether he wants it or not. Consider the surgeon: in no other social context would a competent adult give his consent to another person to stick a knife in him. In no other social context would such an action be met with anything but outrage. And yet not only do we trust the medical reasons given to us for the surgery, not only do we pay our surgeons (indirectly in a public health service) huge amounts of money to do it, but we are even grateful for their knife-wielding skills. Where does the surgeon get the confidence from? What on earth gives the surgeon *the right* to use his knife even on someone who consents? This will seem a strange question to some. I don't mean a legal right based on his elaborate skill and knowledge and ultimately on his certification by the appropriate regulatory body. I don't even mean an ethical right based on the patient's consent or on the likely benefits which will accrue to the patient. Rather, I mean an almost metaphysical right that elevates the surgeon above all other classes of people. This is part of the extraordinariness of medicine. He can do things with his knife that nobody else knows how to do, and he can use this to perform the most primitive miracle of all, that of directly saving life. Consider the hospital: in no other single building in human society is there such an overwhelming concentration of suffering, despair and death. Our normal encounters with suffering and death are piecemeal: an elderly relative gets cancer, the news describes how two people die in a bus crash. There is time to deal with it, time to distance oneself from it, time to move on, more or less successfully. And yet the hospital staff have to deal with one illness after another after another, in the knowledge that there will always be more to come, *forever*. Certainly there is cause for joy after a successful treatment; but this cannot dispel the sheer mind-numbing mass of suffering that they are unable to treat.

(This last paragraph will reveal my own sheltered background, writing as I do in one of the wealthiest and most peaceful parts of the world. For the overwhelming concentration of suffering in the hospital is similar to that of the battlefield and the slum, neither of which are as directly visible to the ordinary person's experience in Britain in 2007. Even our prisons, which are designed to make people suffer, are much more humane in comparison with past forms of punishment. But I think everything I say about the extraordinariness of medicine can probably be said about slums and battlefields.)

This is not supposed to be a banal paean to the heroics of the medical profession. My aim, rather, is to remind the reader of that first awe and horror that he felt as an unprejudiced child upon realising what medicine was all about, and before accepting the story that it was a job just like any other. That awe and horror are the natural responses to the socially extraordinary nature of medicine, and once we are sufficiently reminded of it, it is enough to overwhelm the poor powers of ordinary ethics. The ethical education that most people receive in childhood equips them well enough for the classroom, the shop, the office, the nightclub, the oil platform, almost everywhere. But the hospital – if one really opens one's eyes to what is going on there – will overwhelm every newcomer, no matter what his age or background. Importantly, it will overwhelm the philosopher too. This overwhelming is ethical and not just psychological. After all, it is tempting to reconceive the experience as one requiring 'mere' psychological fortitude and objectivity, the sort of thing required by bungee jumpers – I put 'mere' in scare quotes because I do not want to imply that it is an easy process to grit one's teeth to the stench and the groans and the gore (just as it is not easy to jump off bridges with an elastic tied to your feet), but in a way it *is* easier than having to deal with being ethically overwhelmed.

What I mean by ethical overwhelming is the sudden realisation that there is *no good reason* why this child in bed 47 is desperately ill while you are healthy. The problem becomes massively ethical when this child is multiplied over and over in the wards across the country. The suffering in a hospital is a terrible reminder of the brute vertiginous contingency of our existence.

Consider the mother of that child who asks the doctor 'why does my child have to suffer so?' In one context she is looking for a causal explanation, that is, a diagnosis. He has developed meningitis: the membranes of his brain and spinal chord have become inflamed due to bacterial infection. Or she might be looking for a history of how he caught the infection, whether this was an accident that could happen to anyone or whether her child had been reckless. But there is also a deeper interpretation of the question: why *this* child, why my child, of all the children in the world? And for that there can be no satisfactory explanation, at least not from the doctor. Being overwhelmed in this sense cannot be relieved by gritting one's teeth or ignoring it.[6]

The ethical, however, should not be seen as comprising only situations of great suffering or great risk of death. Moral philosophers are perversely keen on such extreme situations in their examples, as if the essence of the subject matter can only achieve sufficient clarity in this way. But the ethical mostly comprises the mundane and small scale, both inside the hospital and out: every expression of gratitude or apology, for example, is ethical. So the overwhelming nature of the ethical experience in a hospital is revealed not only by the child's reasonless suffering, but also by things like the mundane trade-off between efficiency and kindness. However well meaning, staff will always be too few. Decisions have to be made at each bedside over how long to stay beyond what is clinically necessary, and over how to extricate oneself politely. These are ethical questions since they have a direct impact on patients who are frightened, vulnerable and lonely. Once again, one is struck by there being no good reason not to spend another five minutes comforting this patient, here and now, whatever the obvious reasons for not spending five minutes more with every patient. Or one is struck by the poverty of this patient, who has no home to go to once we discharge him, and whom we suspect of harming himself deliberately for a warm bed and lots of attention. Why him? Why have I never reached that state of despair and degradation?

With the above in mind, consider the problem of teaching medical ethics in the classroom. A typical medical ethics seminar will turn on '-isms' (such as utilitarianism) or 'issues' (such as euthanasia). The sessions might well be popular and generate enthusiastic discussion. They might lead to extensive research and excellent essays. But if the students have not yet been ethically overwhelmed by the medical world, I am not sure that these philosophical discussions can help them make sense of what is really going on. Even when classroom discussions focus on the specific detail of a particular case, the individual is still described in general terms, as part of a search for a consistent approach to types of patient. The intellectual activity of arguing for a specific euthanasia policy is radically different from the stunning intellectual and emotional perplexity of *facing* a particular patient who asks for your help to die. The student may well come to his own conclusion on the matter in the classroom, but such a conclusion is unlikely to bear any relation to what the student will end up thinking after such an encounter. His ignorance is not only of the medical world, but it is also of himself.

The two most important aspects of the encounter are its *particularity* and its *proximity*. Particularity means that all the details of the case are in principle available here, and we can go back as often as we need to. 'Going back' here involves not only a search for further relevant information about the patient's unique situation and wishes, but also the opportunity to talk over the situation with the patient, and help him to discover what his wishes are in the first place: it is too easy to avoid discussion by hiding behind the concept of autonomy and giving the patient what he wants. Proximity means that the patient is 'in your face' rather than summarised in a textbook or on a Power Point slide. There is no avoiding their pain and their anger. Crucially, their proximity means that you will learn something of their point of view. Learning about another's point of view is not a matter of accepting *that* the patient has another point of view – of course he does; rather it is being *struck* by the other's point of view.

Philosophical discussions of '-isms' and issues teach the student to talk the talk without any guarantee that he will properly *adopt* the words. Certainly he will *absorb* expressions like 'autonomy' and 'best interests' and 'quality of life' without really understanding what they mean; but ethical maturation involves adopting the words, being present in them, standing behind them without using them as a shield: 'quality of life' only means something when it is used in a context of a discussion with a real patient making terrible decisions that will affect his quality of life. Does he want to start the chemo now or does he want to wait a little longer so that he can finish a project at work? The particularity and proximity of such a patient making such a decision is what adds flesh to the words, and brings them to life, and it is the health-care professional who comes to understand this better than the philosopher precisely because of his presence at the bedside.

And yet there are two key interrelated objections here. First, it is a notorious truth that in order to remain effective in the medical world in the short and long term, a doctor simply cannot allow himself to be ethically overwhelmed. He has to learn to harden his heart and get the job done. It could thus be said that a doctor should be the *last* person we would ask for advice in an issue of medical ethics, *particularly* an experienced doctor who has hardened his heart that much longer. What the doctor needs is precisely some rough-and-ready ethical principles because he can no longer trust his intuitions after suppressing them for so long. If this is the case, then surely the philosopher's ignorance and innocence has a place (on the clinical ethics committee, say, or in health policy debates) in bringing the very lucidity which I claimed ought to be the goal of philosophical enquiry?

The second objection runs as follows: I have been claiming that the essence of ethics are ordinary (non-medical) situations, using ordinary concepts, and yet I seem to be also suggesting that the medical world is sufficiently extraordinary that ordinary ethics run out. That seems to imply that anything goes, and one might as well flip a coin to decide whether this patient will live or die.

I do not have a decisive rebuttal to either objection and can only acknowledge that they will remain as simmering tensions throughout the remainder of this book. In response to the first, I would simply try to invoke the difference between a good and bad doctor. Certainly there are many doctors who have hardened their hearts to the point of making their work mindlessly routine. While they are not mindless enough to be negligent in their duties, it may be true that the patient's individual worries and fears leave them cold, and that ethical considerations no longer speak to them with the urgency that they should. But is it so naïve to think that other doctors continue to be guided by a deep altruism and humility, and are moved by a genuine concern for their patients' well-being, and that they are able to sustain a commitment to serious ethical reflection about everything they do? This then becomes a new question of how to select and train the doctors of tomorrow.

With regard to the second objection, I will admit that it is a struggle for an ordinary person to make sense of the medical world, and part of that struggle is in the new application of his ordinary ethical concepts to these radically unfamiliar situations. But my point is that his ordinary ethical concepts are all that he has to go on, all he has to start with, and this is what should be recognised by moral philosophers. The nature of the struggle is such that it might not succeed, and the person will be driven to despair or to false oversimplifications, but that is a risk whenever we try to make sense of new situations. It is only with a requisite seriousness and caution that we have a chance of gaining some sort of deeper understanding.

Part II Matters of Birth and Life

The aim of the first part was to reveal something of the distinctive complexity of the problems and disagreements in medical ethics and the inadequacies of the mainstream approach in doing justice to them, let alone trying to solve them. In so doing I adumbrated an alternative account, which I now want to start developing. An essential feature of my account is a careful examination of the place and role of pregnancy and birth in the life of an individual woman and in society more generally, where 'place and role' is to be understood not in social scientific terms so much as in terms of the wider meanings at stake. This is Chapter 3. Only with such an examination of *normal* pregnancy in place can one then proceed to examine the wider meanings involved with the intention to abort a pregnancy in abnormal conditions. One reason to abort a pregnancy might be that the foetus is defective in some way, and I examine this claim in Chapter 4. Three mainstream versions of the claim are criticised on the grounds that they fail to distinguish between the population perspective of the matter and the individual perspective of particular agents involved in the matter – especially the unique perspectives of the pregnant woman and of the foetus. The possibility of the foetus having a perspective then requires an 'argument from potential', and I expand this in Chapter 5, alongside a critique of some of the other arguments used to justify abortion. Note that my aim has not been to come down unambiguously on one 'side' of the debate or the other, but merely to reveal some of the irreducible complexity of the matter.

This argument from potential is based on a wider understanding of what it means for a human being to lead a life through time, and to have a perspective on that life, and this is then the subject of Chapter 6. This chapter also serves as a bridge between Parts II and III, between the broad topics of birth and death.

Part II Matters of Birth and Life

3
The Place of Pregnancy and Birth in Human Lives

The birth of a child is often described as a miracle. The recovery of a seriously ill person, or the survival of a single person among dozens of dead in a train crash, is also described as miraculous. Is this no more than a sentimental flourish to an expression of amazed relief? Can the word therefore be ignored – except when used by a devoutly religious person? I believe the word can be taken seriously, that is, can be taken as substantively meaningful even when used by non-believers, and that possibility is what I want to explore in the first section of this chapter. The problem is not so much about trying to identify what the word 'miracle' refers to, but rather of how the concept is used to capture some aspect of the user's *experience* of the birth of a child beyond the expression of an attitude. This will lead to a more general discussion of the important place of pregnancy and birth in human lives.

One definition of a miracle might be: a welcome event of fateful significance, normally involving an apparent suspension of the known laws of nature. The words 'apparent' and 'known' here already reveal the force of Hume's sceptical discussion[1] and the success of science in explaining away many events previously considered miraculous. And while there are many mysteries left in science, there are no longer any miracles. Obstetrics is a medico-scientific discipline of considerable sophistication that seeks to understand the patterns and causalities within normal and abnormal gestational development and birth. As with any applied science, such an understanding allows better prediction, control and safety of the process, and the field has certainly improved remarkably in these areas over the last hundred years. There are still areas of mystery and doubt, and of disagreement among obstetricians both in the operating theatre and at academic conferences, but these mysteries and disagreement only make sense against a widespread agreement about the relevant 'laws of nature' and about the possibility of further discovery. In itself, birth is probably not mysterious to the obstetrician, let alone miraculous, even if some aspects of a particular birth might be mysterious.

Another definition of a miracle might involve an event that is scientifically explainable but of such a statistically low probability as to invite amazement. But of course birth is hardly statistically unusual. Any large Western hospital will have several dozen per day, most women above the age of 40 have given birth at least once, we celebrate our own and others' birthdays. Birth is everywhere – what is so miraculous about it?

Birth is indeed miraculous, I will argue, in that it seems to involve creation *ex nihilo* of another human being, which contravenes our background understanding of other human beings as permanent. The miracle lies not only in the event itself, but also in the significance of the event as a contravention. Let me explain.

The key to the argument lies in a distinction between two different conceptions of time (and of time passing): one, of time as absolute or objective, and the other, of time as experienced or subjective. The first conception, more familiar to biologists and biographers, describes time as a dimension within which a series of discrete events take place, such that event *C* follows event *B*, which follows event *A*. A person's life can be described as a series of such events, beginning with birth and ending with death. The biologist, like the biographer, can then look for causal relations within the series in order to tell a story of what happened. Within this conception of time, there is clearly nothing permanent about human beings.

However, the story told by the biologist or the biographer is not the story lived by their subject. The obvious point is that the subject is *in medias res*, in the middle of a story, and a story that is *his*: most of the time, the future is essentially open, a lot of it is up to him, and the end is nowhere in sight. In Chapter 6, I will be discussing this notion of living a life 'from the inside' in greater detail. For the moment, the less obvious point is that a person's experience of the present (of time passing) presupposes encounters with other people who are essentially permanent fixtures of the world. I mean this in a very specific way. When I say goodbye to someone, I take it for granted that they will continue to exist, relatively unaltered, and that I can always find them again if I want to. There may be contingent impediments to my finding them again (lack of money, prison walls, etc.), but I can still imagine overcoming such impediments and tracking the person down eventually. This implicit expectation is more tellingly revealed in the common

terms for 'goodbye' in other European languages, which usually trans-
late as 'until we meet again': *au revoir, auf Wiedersehen, arrivederci*. The
fact that a human being is less permanent than a building or a mountain is irrelevant here; what matters is that they are more permanent
than most of my concerns of the present. The fact that a given human
being will have his *im*permanence horribly demonstrated in an auto-
mobile accident after I bid farewell to him is also irrelevant to the spontaneous *expectation* I have of being able to see him again if I want to.
(Clearly I am assuming a context of relative peace, affluence and good
health, without an ever-present risk of death.)

This essential permanence stretches forward and back from the present
encounter. Not only is there a background assumption that the other will
continue to exist indefinitely into the future, but also that the other has
existed indefinitely *until now*. Again, the fact that the person did not exist
200 years ago, or even perhaps 20 years ago, is irrelevant here; what matters is that – if he is now an adult – he was around long before most of
my present concerns, probably with much the same appearance and
character as he has today. Our encounter will be full of references to
shared experiences in the recent past, as well as plans for the near future.
By 'present concern' I understand nothing more complicated than the
sort of things that we worry about, both forward- and backward-oriented
things, in the short term: where can I get a cheap haircut? Who took my
stapler? What's on the telly tonight? How did England do in the last
World Cup? Many of these concerns derive their sense as relatively imper-
manent precisely within the ongoing relationships with other people
(friends, colleagues, shopkeepers) who are permanent.

So the miraculousness of birth derives, at least in part, from the dis-
sonance with the permanence we spontaneously expect all other
human beings to have in our ordinary encounters with them. Here is a
photograph of my (adult) friend Jimbo in 2005, together with his family. Here is another photograph of Jimbo's family in 2003, but without
Jimbo. It makes sense for me to then ask Jimbo where he was, and for
Jimbo to reply to the effect of: 'I had a job in Indonesia at the time'. It
would not make sense for Jimbo to reply 'I hadn't been born yet'. And
even if Jimbo were one year old in 2005, there is nevertheless something
uncanny of seeing Jimbo's family in the 2003 picture without any sign
of the infant Jimbo at all, just as there would be something haunting
about a 2007 family picture that did not feature Jimbo because he had
died in the meantime.

Let me put the point another way. Corresponding to this distinction
between absolute time and experienced time is the distinction I drew in

Part I between what I called propositional knowledge and experiential knowledge, between what we know intellectually and what we feel. I can know that 'all human beings are mortal', that 'Socrates is a human being', but I do not *really* know (experientially) that Socrates is mortal when he stands before me in the marketplace in evidently rude health. In other words, my propositional knowledge of his and others' mortality does not undermine my assumption of his permanence (in relation to my short-term concerns). And so it is, going backwards as well, with birth and childhood. We know that 'all human beings are born' but that propositional knowledge does not enter into the encounter with Socrates here and now, nor does it enter into the ongoing relationship that I have been having with Socrates until now and plan to continue indefinitely. Socrates seems always to have been Socrates. Of course I can ask him about his birth, when it was, where it was etc., but this is not a conversation about *Socrates*, but about the infant who would become Socrates.

This notion of a sudden appearance, then, is part of what makes the birth of a human being miraculous. But there is more to it than that, for the human being does not just appear but *is born*, that is, born from another human being. And it is this notion of creation that makes the event truly miraculous.

Once again, there will be plenty of intellectual knowledge, lay and obstetric, about how children are created, and no adult today will deny the truth of the biological story. But at an intuitive level, I suggest that the biological story is very difficult to endorse when *confronted* by the startling conclusion of that story. As the mother cradles the newborn infant in her arms, there is no way that mere biology can amount to *this* – this squirming, screaming pink bundle of human life. Nine months ago this person was not here and, indeed, not anywhere. Adult humans are used to secreting and excreting various things from their bodies, but such things are all undifferentiated matter; matter that can be wiped and washed away without much thought. But to squeeze out another human being is utterly exceptional, surely, an experience that bears no resemblance at all to any other human experience.

Yes, in principle the biological constituents already existed, the atoms and molecules were 'waiting' to be assembled according to the DNA blueprint. And even if the process is extremely complicated, the biologists are justifiably confident of making progress in the direction of greater understanding. And yet there remains an abyss between this *type* of understanding and the type of struggle required to make sense of the outrageous fact of a new human being. Not only does birth seem to defy the known laws of nature, then, but the birth of *this* child also seems to be an event of gross statistical improbability – part of our second definition of 'miracle'. No matter how many other children were born at the same moment, these will be utterly irrelevant to the parents' experience of their child in all its uniqueness. Of all the human beings that could have been born, of all the sperm that could have encountered the egg, of all the men who could have met and impregnated the mother, what was the infinitesimal probability that precisely this human being would come into the world? (Although 'probability', a concept more at home in the natural sciences, may not be the best word to characterise the astonishment in the face of something *super*natural.)

The interesting point here is that the rest of us are not tempted to reproach the parents for such astonishment. The midwife does not scold them for projecting superstitious significance onto a mere fact. The parents are not 'blind' to some scientific 'reality'. On the contrary, I would suggest that birth is a very pure example of what it means to ordinarily acknowledge another as a unique human being.

Creation and flesh

If birth is indeed supernatural, even for the non-believers, then one has to ask what role the parents play exactly in the creative process. As elsewhere in this book, I want to start by looking at the meaning and resonances of the concept of 'creation' in its ordinary or paradigmatic uses. The carpenter creates a table, for example. He looks around his workshop at the various bits of wood, he selects the tools he will need, he draws a quick sketch of what he has in mind, all the time guided by an *idea* of the table at which he is aiming. This idea might be quite precise, and he will be satisfied if he only approximates it; or the idea might be quite vague and would only acquire sharper contours as work progresses. But either way, given the function of a table he will know whether he has got it right or not, whether it is a sufficiently good table for the customer's purposes (and pocketbook). Most importantly, the carpenter is very clearly the author of the resulting table: it was his idea, he did the work, and it was not an accident that a table was produced.

I want to argue that the parents cannot be called the creators or the authors of the child in this paradigmatic sense. They have no knowledge of components or tools, they only know about a single necessary trigger mechanism, their knowledge of the biology is probably sketchy at best, and they have no clue whatsoever about the precise properties of the eventual product. They know that a certain type of being may result in nine months' time, but they know nothing about the unique nature of that human being, i.e. little Jimmy, with his curl of blond hair etc. And even if their idea of the product is as vague as the carpenter's might be, the parents cannot *fashion* the human being in the way the carpenter can fashion the table, every step of the way, adjusting his method and material as the result slowly takes shape. All the parents have at their disposal is the trigger mechanism. The parents are going through the motions, so to speak, and then it is nature that takes its course, and nobody knows what will be the exact product, let alone what the infant will turn into in 30 or 60 years' time.

But here is a subtle distinction. The parents cannot claim authorship, but they can claim that the child is *theirs*. Clearly this is not a matter of legal ownership (in the sense that the house or the car is theirs), so where does this use of the possessive article come from? In the long run, it has a lot to do with the intimate identification with the growing child, in a similar way that a stranger becomes 'my' friend. But what is the basis for the possessive article in the case of the newborn? In what follows, I want to concentrate on the mother, and will be arguing that the child is 'hers' because it is *of her flesh*; and that this is to be contrasted to the table's relation to the carpenter. This is a word that is much more popular in the phenomenological tradition than in English-speaking philosophy, so I have to tread carefully to avoid obscurity.

One way to understand the foetus as flesh is to contrast it with a different picture, which takes the foetus as essentially a separate object from the mother, essentially detachable. According to the textbooks, the sperm meets the egg in the uterine cavity, the resulting zygote then implants in the uterine wall. There an umbilical cord develops to supply nutrients to and remove waste from the growing embryo, and later, the foetus. So far so good. But note that such a description makes no necessary reference to the woman: in principle, there is no reason why the sperm could not meet the egg in a test-tube (as indeed happens in these days of IVF), and no reason, in principle, why the embryo and foetus need to develop *in a woman*; it is merely a question of time, some say, before antenatal life-support technology improves sufficiently to provide all the foetus's needs, and the discomfort of pregnancy and the agony of childbirth will become things of the past. Throughout, the growing foetus is already seen in terms of the separate individual it will become, and the woman (and her digestive, respiratory and immune systems) provides the most effective temporary support.

However, I suggest that this picture does not correspond to the experience of most women. It is not that they would disagree with the biological detail of the picture; rather, it is that the business of pregnancy and birth carries a different significance in their lives, with different implications for the relationship with the child, after the birth. For a start, the foetus, especially in the earlier stages, is not experienced as essentially separate and detachable; even after artificial fertilisation and insertion, it is not therefore a foreign body like a pacemaker or a contraceptive implant. Rather, the foetus *grows from her*. It is not as if she eats food and breathes air in the knowledge that she is merely channelling these 'biochemical building blocks' into the umbilical cord for the pre-programmed assembly of a separate organism; rather, she is eating and breathing for herself, and the foetus is part of herself. Birth is not a matter of opening up the boot and removing the luggage. It is a bloody, messy, painful wrenching of the baby from the viscera of the woman.

I realise this sort of language will be off-putting to philosophers and scientists alike, perhaps because they have become enchanted by the embryological, cellular or even molecular descriptions of what is going on. This cell whizzes over here and combines with that one, the DNA is replicated, the heart first develops as a thickening tube etc. After all, there are two banal senses in which the foetus is the same flesh as the mother: they share the same micro and macro structure of the biological species of *Homo sapiens* (so that we are all of the same flesh) or they share a sufficient amount of DNA to be related. But these descriptions are completely unhelpful in trying to make sense of what the woman is experiencing: first, her child, and her relationship to it, are entirely different from all other children and her relationship to them; second, her experience of pregnancy and birth does not include an experience of DNA, for she has never seen DNA, let alone *her* DNA. I am talking about the sort of experience that women had even before DNA was discovered.

My point might be misunderstood in a different way. Surely I should make my mind up, it will be argued: is the foetus part of the woman or is it a separate human being? Answering this question is often taken as crucial for one version of the abortion debate: after all, if the foetus *is* part of the woman, then the woman is entitled to have it removed, like an appendix or a tumour, without any moral repercussions. Although I will leave a more detailed discussion of this topic to Chapter 5, I would reject the dichotomy as false: the foetus is *both* part of the mother and separate. The happily pregnant woman is not talking to her belly, but to her child, very much a separate being, despite being still *in utero* and entirely dependent. And yet, it is inside her, very much part of her, growing from her.

Perhaps the most controversial part of the picture I am developing is the claim that the child, after birth and after the cord is cut, *continues* to be of the flesh of its mother; and that it is this fact that accounts for the special use of the possessive pronoun by the mother. Even when the child is fully grown, the mother can still look at him and say to herself: 'You were not only inside me, you grew from me, you are made of the same stuff as I am. We are one flesh'. Or we might put it this way: there is a physical link between the two; that it is in the past in no way compromises that fact.

This will seem either absurd, or some bizarre Freudian persecution fantasy. There is no physical link there, and so mother and child are just as independent as any other two human beings on the planet. If they are both adults, they can vote separately, own separate property, sue and be sued by each other etc. Insofar as there is a link there, it is merely a contingent emotional bond, one that could in principle develop between any two human beings, but which is perhaps more likely to develop between mother and child in virtue of the many years of intimacy. Again, it is not that I want to deny the above picture, nor the importance of the many years of intimacy, nor indeed the importance of facial resemblance that begins in childhood (the closest that the genetic story gets to the surface). All I can do, perhaps lamely, is suggest that such factors are not quite enough to account for the full *meaning* of the woman's experience of pregnancy, birth and motherhood. It is one thing to describe, from the outside, the intimate bond between mother and child and speculate about its causes; it is another to try to make sense of the experience, from the inside. Note that my account is still compatible with the mother not liking the child, and even going on to abuse it.

My account also suggests that the father can never have the same sort of relationship with his child precisely because it is not his flesh; he can only watch and envy the mother-child bond. Despite the fifty-fifty division in the genetic story, the father's experience of pregnancy after his modest physical contribution all those months ago is that of a mere onlooker. Again, he certainly does not deny the genetic story, but it remains too abstract for him to identify with the swelling in his partner's belly. Nor does he deny that the grainy image on the ultrasound is his child. But he does not have anything like the experience, for better or for worse, of the child growing inside him and eventually being wrenched from his viscera. Neither of the parents sees the child until it emerges; but the woman can feel it kicking, and this uncanny sensation – which would be horrific and revolting in any other context[2] – yields a very different kind of knowledge to that acquired from seeing the ultrasound. (Two comments: it has always struck me as odd that babies are named after fathers. Surely if it comes out of the mother, it should be named after the mother. Perhaps originally it was a way of reassuring a father about his paternity when families and dynasties were much more important. Second, the father's subordinate role during the pregnancy and birth does not let him off the hook when it comes to the business of raising the child, of course.)

The problem with descriptions

With the above in mind, I now want to pick up my brief discussion of personhood from Chapter 1, in order to expand it within the context of the abortion debate by looking at the descriptive concept of moral status. Chapter 5 will then contain a more sustained and detailed discussion of abortion.[3]

One way of framing the abortion debate is to focus on the so-called moral status of the foetus. At one extreme are those who claim that the foetus, even at an early embryonic stage, has the same moral status as an infant, and with it the same moral rights and protections. This position is adopted despite the radical dissimilarity between the early-stage embryo and an infant, in terms of its external appearance, its internal organogenesis and its psychological capacities. At the other extreme, the foetus is taken to be no more than a bundle of cells, akin to an appendix or a tumour, with no moral status at all and fully the property of the woman to have it removed if she wants to. Between the two extremes are various 'moderate' positions, which can accept the gradual change in the moral status as the foetus comes to resemble the infant more and more, usually in virtue of the development of a key property; abortion then becomes permissible at the early stages and impermissible in the later stages.

This much I am taking as familiar. What is uncontroversial for most parties seems to be the belief that the object in question can only have a single, correct empirical description at any given time, and that its moral properties will somehow follow self-evidently from this description. The problem begins when we recognise how many descriptions *every* object has, depending on the context within which somebody feels the need to describe it. After all, a plastic bottle is both a useful container and a piece of litter. Recall the disagreement between the Vegetarian and the Carnivore in Chapter 2. Yes, there is a physical–biological description, but this cannot amount to the full meaning of the word 'rabbit', *within human practices.* The physical description can be used when trying to distinguish one object from different ones around it ('can you see the rabbit in the picture?'), but it cannot capture the ways we interact with the rabbit, many of which are more than just observational. For in different contexts a rabbit can be a pet, a pest or a pâté, and thus calls for different actions as appropriate in different contexts: to stroke it, to shoot it or to eat it. And yet the hapless rabbit is the *same* rabbit throughout, just doing what he does naturally – he doesn't deserve the place that he has acquired in the various human practices that concern him, nor is it possible to speak of it as being 'really' a pest. Finally, of course, the rabbit is a 'bunch of cells', just like the human foetus whose status is so much in question. But when the human owner of the rabbit pet strokes its fur, is he stroking a 'bunch of cells'? He is and he isn't: at one level of description both owner and pet are bunches of cells, but does such a description say anything interesting?[4] What is more important, surely, is that the owner, within his perspective of the activity, is *not* stroking a 'bunch of cells'; and when the hunter brings the rabbit back for the evening's stew, his family are not eating a 'bunch of cells'.

My point about the difficulty in finding a description that is sufficiently neutral over multiple contexts even applies to certain *names* as well. While 'rabbit' is neutral enough to be the agreed name in a dispute between vegetarians and carnivores, the name 'foetus' cannot claim such a privilege, despite the apparent neutrality afforded by its use within scientific medicine. Scientific medicine is but one context in which the foetus exists, and there is *no reason* to think that it should somehow be the dominant or foundational context. To put the point another way: in most discussions of abortion, there is not enough attention paid to the fact that it is an artificial intervention, used when something goes wrong. The natural and normal thing to happen is for a woman to want to get pregnant and to want to stay pregnant: this is the paradigm, and *this* is where we should be looking for the moral status of the foetus. Consider the happily pregnant woman, stroking her bulging belly and whispering lovingly. Within her perspective, she will not be whispering to her 'foetus'; she will not tell her sister that her 'foetus' is coming along nicely. She will instead speak of her 'child' or her 'baby'. This is not just a lay term, while the object that she is referring to is 'in fact' a foetus all along; this is not a baby 'for her', while it is a foetus 'for everyone else'; this is not a question of levels of description, such that the cultural name or shorthand or sentimental nickname supervenes on the scientific name. No, it is not a foetus at all while she is stroking her belly and whispering to it, that is, while she is *engaged* with it. Nor will she admit later that she was just being sentimental and say 'yes, of course it was a foetus, all those hormones distort your perception' – her experience was of her baby, and her memory is of her experience with the baby. In saying that this is the paradigm situation, it places the burden of proof on others to somehow demonstrate that the being in question is not a baby. Yes, of course the object might be both foetus and baby; but using the word 'foetus' as somehow foundational already allows the possibility that it is *not* a baby. Merely pointing at empirical features is not going to sort this matter out either way.

The same woman then goes for an ultrasound scan, and the sonographer *might* say that the foetus 'is coming along nicely' (although chances are that the sonographer would speak of the 'baby'). And the mother knows what the sonographer means, she doesn't deny that what he is referring to is a foetus. But that sort of conversation takes place in a highly specific scientific context. A trained technician is looking for developmental defects in an organism, so of course he will use the language of science. But this is not because the language of science is more accurate. It is because of certain distinctions that are important to scientific study and scientific instruments, and the sonographer comes to see the object with his task in mind. The pregnant woman is not interested in those distinctions while she is stroking her belly, for she has come to see the object in a very different way. To such a woman the suggestion that the foetus is only a 'bunch of cells', with the moral status of other removable bunches of cells (such as an appendix) will seem unintelligible, even monstrous. The implication is that if she were to lose the baby through a natural miscarriage she would not really be losing anything more than an appendix; this would suggest that women who were *upset* by their miscarriage, or who *grieved* for their lost baby, were being sentimental or confused.

Now compare the happily pregnant woman's name for what is inside her with the word 'foetus' in the following sentence, taken from a popular textbook on medical law and ethics:

> in carrying out an abortion, the doctor would be wise to ensure that the foetus is killed while it is still inside the uterus. If the foetus is alive outside the uterus it may acquire the legal protection of any newborn baby.
>
> (Hope, Savulescu and Hendrick 2003 p. 122)

The fact that the second sentence might well be true, and that on the whole the advice might be sensible for an obstetrician to follow, seems to be irrelevant to the chilling callousness of the description. The question of the foetus's moral status *is already settled* here – partly by the very use of the word 'foetus'; there is no room for ethical doubts, and all we are left with is the practical question of protecting the well-meaning doctor from the arbitrary vicissitudes of the clumsy law. Clearly there is something very strange going on here if mere physical location relative to the uterus makes such a difference to ethical and legal status. But more than that, the whole *tone* of the advice sounds like there is something to be covered up.

If I want to argue that the lawyer is blind to the moral reality of the foetus, then I might need to understand *how* the lawyer settled the question in the first place; how he 'came to see' the foetus, incorrectly, as the sort of being that it would be wise to kill before it acquired legal protection. The whole point of the dialogue between the Carnivore and the Vegetarian in Chapter 2 was to show that people bring their 'ways of seeing' to the situation, and if they meet people there who see the object in the same way as they do, then (as in this case) there may be no ethical problem, no question of moral status to settle, merely a practical problem, in the context of which the lawyer's advice to the doctor is eminently sensible. If I, on the other hand, felt the doctor and the lawyer were being callous in 'making sure', then I face the problem not of persuading them but of bringing them to see that what they are killing is – is what? Is a baby? How do I do that? For there are probably no *facts* about the object that I can tell them that they do not know already. And the ethical disagreement jams.

To see the foetus as a baby seems to imply a belief in the sanctity of its life. And the concept of sanctity is also invoked in the euthanasia debate as well. Given that this chapter is laying the groundwork for later, more issue-specific chapters, it is worth briefly exploring the concept in order to see how much more complicated it is than it at first appears, and how it relates to pregnancy and birth. It is not always clear exactly what sort of work the concept is supposed to do in these debates. As with miracles, sanctity is a concept with obvious religious significance, but one that is also used by non-believers in a context with no overt link to organised religion. I want to take such use by non-believers seriously, and explore the implications of sincerely describing something as sacred. (Those who are bothered by the religious implications tend to use the word 'inviolable'.)

The original notion of the sanctity of a human being's life was logically tied to thoughts about God's role as creator and sustainer. Although God created all the world and all living things, humans were especially sacred because they had been created in His image. And because it was God who created humans, He was the only being with the ethical authority to bring their lives to an end. This is one aspect of the meaning of sanctity. Another aspect has to do with my sharing the status of God's creation with all other human beings, that is, being equally vulnerable to the same kinds of harm and humiliation and bad luck, and above all being equally mortal. To recognise another's life as sacred is therefore to recognise that 'there, but for the grace of God, go I' – which is possibly the most profound truth in ethics. It is also the root of the compassionate response to another's incomprehensible and unjustifiable suffering in the medical world. As Raimond Gaita (2004 Chapter 3) describes it, this common vulnerability is a far more plausible candidate to capture the essence of human beings than the favourite category of modern moral philosophy, rationality. It is hard to imagine standing at the bedside of a dying patient, and reflecting on the rationality that we both share!

This is therefore what I propose as a secular interpretation of the term 'sanctity'. The problem in the abortion and euthanasia debates is that the concept of sanctity cannot support a clear argument; for instead it is a way of seeing. Merely shouting 'but it's sacred' at your opponent is not going to persuade him. Rather, the challenge is to bring the other to see that all human life is sacred. I stress, however, that in the paradigm situation of the happily pregnant woman, there is no *need* to invoke the sacred. Within the woman's perspective, the 'baby' is obviously sacred, about as sacred as anything could be in that woman's life, especially given its enormous vulnerability.

Let me conclude this section by asking a blunt question: does this mean that I oppose abortion? It is a difficult question to give an unambiguous answer to, for reasons that will become clearer in the next chapters. For the moment I will say this much: treating the life of another human being as sacred – recognising the sanctity of their life – is compatible in certain extreme circumstances with aborting that life, and indeed compatible with withholding life-saving treatment (from neonates, from the comatose etc.). Perhaps less controversially, it is also compatible with finding it necessary to protect oneself or one's loved ones against a credible threat by using proportionate defensive force that may carry the risk of killing the aggressor. The key word in this last sentence is 'necessary', rather than 'justified', and here I am recalling our discussion of moral incapacity in Chapter 2. The whole point of taking something to be sacred would be lost if it were thought of as subject to qualification and justifiable exceptions. The concept of sanctity of human life would merely become a bland general statement to the effect of 'human life is sacred, except for circumstances *A*, *B* and *C*'; and then the haggle would begin over the exact nature of those circumstances. Once the exception is justified, however, then that is the end of the matter. But if you consider life to be sacred, absolutely sacred, then killing it will always be wrong – even though one might find it necessary. The difference lies in the mindset with which one confronts the situation, and in which one then walks away from the situation.

Learning to love

The main purpose of this chapter has been to advance some considerations towards a more nuanced understanding of the place of pregnancy, birth and childhood in human lives, and thereby to set the scene for later discussions of certain issues in medical ethics such as abortion and the withholding of treatment from handicapped neonates. With this aim in mind, I now want to introduce a third scenario. Rather than the paradigm of the happily pregnant woman, rather than a woman who considers life sacred but who nevertheless finds she has to go through with an abortion, I will consider a woman who is ambivalent about an unintended pregnancy but learns to love the child she eventually bears. What interests me are two things here: first, how she comes to see the 'pregnancy' as a 'child', as sacred, and indeed as *her* child; second, how she accommodates (if that is the best word) her emotional life to the new arrival. After all, love is not normally something one can just switch on at will; and learning to love sounds decidedly contrived, if not self-deceiving.

There is a double ambiguity about the concept of 'wanting' in the context of making decisions in pregnancy: (i) an ambiguity about whether the concept of wanting is even applicable to certain pregnancies and (ii) an ambiguity about whether one can want something that one can barely imagine. Let me consider these two ambiguities in turn.

The concept of 'wanting' has its natural home in the supermarket. 'Do you want beans or chips for dinner tonight?' I ask my friend. He chooses from the shelf accordingly. If he changes his mind later, he certainly doesn't have to eat the beans – he can return the tin or give it to someone else or leave it for later, all on the basis of his changing and knowable wants of the moment. Nothing mysterious about this.

Now let us take this pregnancy that is unplanned and inconvenient for various reasons (even if one cannot say it was completely unexpected despite the best of precautions). Consider the following possible reaction: rather than asking themselves whether they 'want' it or not, the couple take it as *given*, and re-organise their lives to accommodate it. For this couple, abortion was never an option. In this way, the pregnancy is initially taken as any other event that is neither wanted nor planned, such that the event's status *as* wanted or unwanted becomes irrelevant after the fact. A similar case might be that of a relative falling seriously ill and requiring care. I deal with it as best I can, according to what my relative needs: it never occurs to me to reflect on whether I wanted it or didn't want it to happen. Of course there are other decisions that I will have to make within the context of the on-going care, and here it does make sense to speak of wanting, e.g. wanting some time off this week-end and hiring a carer to cover my absence. And of course I may spontaneously resent my relative for falling ill just now, when things were going so well in my job. But such resentment is not the same as 'not wanting' because it does not get me anywhere further in my thoughts or deeds. Normally 'not wanting' is a response to a perceived choice. I quickly realise that it is hardly my relative's fault, that I cannot do anything about the illness, that I have to continue caring for him, and so I refrain from endorsing the resentment.

However, the unplanned pregnancy is importantly different from my relative's illness. There might not be very much room for seeing the 'bright' side of my relative's illness while he is suffering, especially if the prognosis is bleak. So I struggle on stoically. But the parents of the unplanned child are concerned to do more than just struggle stoically, for they realise that the stoic attitude may well undermine any potential that they – or more importantly, their child – have for a happy and successful family life. The child, they say, must never be allowed to suspect that it was not initially wanted, for it might then suspect that it was still unwanted. And so with an initial effort they manage to accommodate the child in their physical lives, but also gradually to accommodate it in their affective lives, as it were. They learn to love it. By the time they have reached this stage, it is as if they have wanted the child *all along*.

This line of reasoning invokes two ideas from existentialist thought: first, the idea of 'committing' oneself to a role or ideal and thereby cultivating, over time, the appropriate attitudes and dispositions; and second, the idea of *amor fati*, the drive, literally, to love one's fate as if one had wanted it and chosen it all along. But this also looks glaringly and dangerously like self-deception. Making the best of an inconvenient situation is one thing, but wilfully re-writing one's past is surely another. In dealing with the sick relative, I resist the temptation to sulk or whinge about it or to daydream about all the other things I could do with the time and the money. Rather, I just don't think about it. I concentrate on the task of the moment and make sure he gets the attention and the treatment that I feel he deserves. So this is not self-deception so much as selective attention.

Before we consider whether 'learning to love' involves self-deception, or at least a pernicious form of self-deception, let me return to the second ambiguity about the concept of 'wanting', that is, to the question of whether one can truly want something that one can barely imagine.

How informed can the decision to get pregnant, or to keep an unexpected pregnancy, be? This links back to my earlier point in Chapter 2 about the smoker's failure to imagine what having lung cancer will be like, and the corresponding difficulty in saying that his consent to smoke is sufficiently informed. Some of us may already have experience with small children: primary school teachers, paediatricians, and those who spend a lot of time looking after nephews or cousins. But for the vast majority of us, having one's first child – the pregnancy, birth and everything afterwards – is a *radical* change to one's life. One can read the books and talk to recent mothers as much as one wants, but the most important things cannot be described so much as discovered on the job. This is partly because every baby, just as every parent's experience of birth and baby, will be different in unpredictable ways. But mainly it is because the experience is something that 'consumes' the experiencer. By 'consume' I mean that the experience is much more than an isolatable set of events upon which I can bring my intellect and will to bear; rather, the experience changes my very character and will, brings me to see myself and the world differently.

Consider the word 'cope'. As a manager of a company or a school, I have to cope with many unpredictable situations, and I will cope more or less badly depending on my skill, experience, colleagues and luck. But if worse comes to worst I can always resign from the job, and say 'I did my best, but it was not to be'. Now consider the 'coping' involved with being a parent. Ahead of time it is hard to predict not only what will happen, but the *relentlessness* of what will happen, and the *exhaustion* that one will feel in response to such relentlessness. One is always surprised, as a new parent, just how many reserves of strength one can call upon, reserves that one would probably not tap for any other commitment. Throughout, there is the thought that 'I will cope because I *have to* cope. I don't know exactly how I will cope, but I know I have to'. I can leave my manager's job when it overwhelms me because I know there are other jobs. But leaving my child is never more than a fanciful notion.

Of course some parents are overwhelmed and give up their child for adoption or have their child taken away from them by Social Services. But unlike the manager's option to quit, the possibility cannot be regarded by the parents as a *viable* option ahead of time, and cannot be remembered as a *chosen* option after the event. Part of the manager's thoughts around his job are the alternatives: 'if they push me too hard I can always do *X*'. There is enough of the manager *outside* the job to form a stable platform from which to choose from the options before him. But the parent is consumed by the experience of parenthood and by the uniqueness of the child; there is no other option until it is forced upon him. The parent can and does make plenty of choices *within* parenthood, but not *about* parenthood. And if the parent reaches the end of the tether and does give the child up for adoption, this event will probably be so traumatic that it, too, will change the parent in profound and unpredictable ways that could never be described as rationally wanted. Parenting can never legitimately be thought of as an experiment that might go wrong, even if it might go wrong.

As with the first ambiguity surrounding the concept of 'wanting', this second one also suggests some sort of self-deception. In the first case I refuse to admit that the pregnancy was inconvenient; in the second I refuse to admit that child-rearing is actually quite tedious, grossly disruptive, and that I don't really like the little brat anyway. But I have learned to love it. Is this self-deception? Perhaps, but I would argue that it is not a pernicious kind. After all, there are certain things that are probably worth deceiving oneself about. To dwell upon the smallness of our planet in the huge universe, for example, could be utterly incapacitating (I am never sure how astronomers preserve their sang-froid). On the other hand, to deceive oneself about one's abilities as a manager is only to bring on one's dismissal as a shocking surprise. Self-deception of the 'small planet' kind is necessary in order to lead any sort of *life* on that planet – there is no rude surprise that one will bring about in the process (unless of course the planet is hit by a huge asteroid, after which one won't be around to worry any more). Self-deception of the pernicious kind involves the assumption that there is a point to facing the truth about oneself, and that point is that one can improve oneself or at least choose a job for which one is better suited; in other words, there is an essential reference to the future. Learning to love one's child is therefore a 'small planet' type of self-deception, since the parents cannot choose an alternative child to live with or an alternative planet to live on.

This discussion of self-deception has been a little cursory, but I will be returning to it at several points in the chapters to come. For the moment, the important thing to realise is that each individual has a life to lead and necessarily experiences the world from within the context of that ongoing life. In order to understand the place of pregnancy, birth and children in that person's life, it is important to make sense of this perspective. Often, my perspective on the world will coincide with a more objective perspective of what is going on, but not always. And many discussions of the difficult issues in medical ethics go astray, I believe, precisely because of the unjustified neglect of this more personal perspective on the matter.

4
The Clash of Perspectives

The two perspectives in question are the individual and the population perspectives. In this chapter, I want to argue that a number of problems in medical ethics can be boiled down to an irreconcilable clash between these two perspectives, a clash that is often 'resolved' by giving priority to the impersonal population one. I will argue that this is an arbitrary resolution, one that fails to make sense of the uniqueness of the individual and the way this individual experiences – that is, from his personal perspective – the dilemmas typical in the medical world. Let me start by taking three different examples from dilemmas associated with pregnancy and birth, from three prominent mainstream authors. My discussion will then follow the third example.

My first example is taken from a piece of recent legislation, the *Human Fertilisation and Embryology Act 1990*,[1] which is designed to regulate assisted reproduction in the UK through the Human Fertilisation and Embryology Authority (HFEA). The Act is interesting to philosophers because it was directly influenced by a report from a committee chaired by a philosopher, Mary Warnock (and published as Warnock [1985]). The Act carefully sets out the conditions under which a couple can be deemed infertile and can request reproductive assistance. (There is a separate ethical question of whether the NHS should pay for the service, which I will not be discussing here.) Importantly, the Act also has a clause about the 'welfare of the child':

A woman shall not be provided with treatment services unless account has been taken of the welfare of any child who may be born as a result of the treatment (including the need of that child for a father), and of any other child who may be affected by the birth.

(Section 13[5])

Warnock later wrote a book (Warnock 2002) about the committee discussions leading up to the passing of the Act, and the above thought was voiced in terms of the 'suitability' or otherwise of the couple applying for the in vitro fertilisation (IVF), and she considered, as a test case, a couple with a proven history of child abuse (p. 45). In the end she and the committee refused to formulate any set of criteria by which to judge social or moral suitability, and left it with the above wording. One reason for her reluctance to formulate criteria was that 'people can and do change'. Such optimism is one of the basic principles underlying our criminal justice system: no matter how many times a competent person may offend, it can never in principle be justified to lock them up for life solely on the basis of a probability about the future (the situation is different for mental health patients).

IVF thus sits uneasily between two paradigms. On the one hand is natural reproduction by sexual intercourse, where parental suitability is almost never invoked as a reason for preventing conception.[2] On the other hand is adoption, where would-be parents are carefully screened to see not only if they are suitable, but whether they are suitable for one type of child rather than another; here 'suitability' has to do with maximising the child's likely welfare. The concern for the welfare of the child in these instances is reflected in other UK legislation.[3] My interest in the IVF example is not in the criteria, nor in the possibility of formulating criteria, nor in whether people can and do change (or whether people should always be presumed to have the ability to change). Instead, it is the idea of rejecting the possible creation of a child on the grounds that its welfare would be in jeopardy.

My second example comes from the debate surrounding pre-implantation genetic diagnosis (PIGD). A number of eggs can be successfully fertilised and then tested for certain genetic defects. The defective embryos can then be discarded, and one or more of the 'healthy' embryos implanted. I put 'healthy' in scare quotes because there is of course no certainty that it will be free of other, presently undetectable defects; but the definition of health in this context can also be controversial. I have in mind those cases where a couple request that a 'defective' embryo be implanted; or, slightly less controversially, who refuse testing and selection even though there is a high probability of an inherited 'defective' trait.[4] In the most celebrated case scenario, a couple who are both congenitally deaf request that only those embryos that will develop the same deafness are to be implanted, and the healthy embryos discarded. The argument was that deafness, in this instance, is a *culture* and not a defect. The deaf community has its own language, its own poetry, its own schools and activities and its own conception of itself as an independent culture that outsiders can never fully understand. In this way deafness was to be compared to any other minority culture such as being Czech outside the Czech Republic or Jewish outside Israel. And just as two Czech or Jewish parents would quite reasonably want their children to grow up Czech or Jewish – no matter what the disadvantages of such a cultural identity – so too deaf parents could legitimately ask for deaf children.[5]

John Harris forcefully argues that deafness is a disability in that it removes the possibility of enjoying certain experiences that most people take for granted. To support his argument, he compares deliberately choosing a 'deaf embryo' over a 'hearing embryo' (the established shorthands in the debate) to the uncontroversially harmful act of deliberately *deafening* an already hearing infant.

> [My critics] sometimes talk as though deafness were not a harm or a deficit. Would the following statement be plausible — would it be anything but a sick joke? 'I have just accidentally deafened your child, it was quite painless and no harm was done so you needn't be concerned or upset!' Or suppose a hospital were to say to a pregnant mother: 'Unless we give you a drug your foetus will become deaf. Since the drug costs £5 and there is no harm in being deaf we see no reason to fund this treatment.' But there is harm in being deaf and we can state what it is.
>
> (Harris 2001 p. 384)

Harris is careful to distinguish his argument from a misinterpretation by his critics. He is certainly not arguing against the possibility of disabled parents procreating *at all*, nor is he saying that the disabled child, *once born*, necessarily has a life that is less valuable than that of a healthy child. Instead, he argues that (i) where there is a choice of what sort of being to bring into the world, it is harmful to choose the more disabled one and that (ii) deafness involves harm, as revealed in the theoretical suggestion of harming a hearing infant.

The third example is taken from the debates surrounding abortion (I return in greater detail to abortion in the next chapter). Glover's position is that of a more thoroughgoing utilitarianism, something that even Harris shies away from.

> [...] it is sometimes wrong to refuse to have an abortion. If tests have established that the foetus is abnormal in a way that will drastically

impair the quality of life, it will normally be wrong of the mother to reject abortion. Most of us would accept that it would be wrong deliberately to conceive a handicapped child rather than a normal one, if the choice were available. On the moral system defended here, this can be extended to abortion. If aborting the abnormal foetus can be followed by having another, normal, one, it will be wrong not to do this.

(Glover 1977 p. 145)

In the immediately subsequent paragraph, Glover qualifies his claim: the abortion should not be carried out if (i) the side-effects are too severe on the woman; (ii) the quality of life of a foetus who is only lightly handicapped (e.g. missing a finger) would not be impaired sufficiently to justify abortion; and/or (iii) the woman might not be able to have further replacement children, and so would be willing to accept this handicapped child. He concludes by saying that he has no 'general formula' to offer (p. 146).

Like Warnock, Glover does not want to spell out criteria for a foetus to meet before it can be deemed abortable. Most of his discussion concerns very severe handicaps, and he is right to distance himself from very mild handicaps. But from what he says later, it seems that he would allow for less severe handicaps as reasons to try for a replacement. 'If someone with a handicap is conceived instead of a normal person, things turn out less well than they might have done.' Of course, once we start speculating about 'less well than they might have done', there is no obvious reason why we shouldn't abort an *average* foetus in favour of one likely to develop above-average talents and skills – this is the debate about 'human enhancement'. Indeed, Glover even goes so far as to compare *himself* to a potentially more talented embryo that his parents might have conceived: 'from an impartial point of view, it would have been better if some more gifted or happier person had been conceived instead' (p. 148). For the purposes of this chapter, I will concentrate on Glover's more plausible claim about severe handicap, but I will return to his comments about himself later.

Individuals and uniqueness

I will be arguing that all three arguments are based on a mistaken understanding of the individual, or more precisely of the uniqueness of that individual, of the individual as a unique subject of experience, and of the individual as leading a unique life. Clearly the central concept here is that of uniqueness, and so it will be worth analysing. Stephen Wilkinson (2003) provides a useful starting point in his discussion of the exploitation allegedly involved in prostitution and purchasing human organs. Part of the wrongness of consensual prostitution, it is generally claimed (following Kant), lies in the failure to acknowledge the uniqueness of the individual whose body one has purchased. In making the purchase, I am treating the person as essentially inter-changeable (fungible) with any other person, since I am only interested in the general features of the physical body, and there are many people with sufficiently similar features. As Wilkinson points out, however, this tends to be what we do in our consensual relations with most people fulfilling roles: waiters, hairdressers, bureaucrats etc. So it is not clear how the prostitute's uniqueness is relevant.

Even if it were relevant, however, it is not clear what it means to be unique. Wilkinson (p. 49 ff.) considers the *empirical* claim that people are in fact different from each other, in terms of both their physical (height, eye colour etc.) and psychological properties (memories, inter-ests etc.). The empirical claim could be true or false, depending on the criterion of relevance and the scope of the enquiry. At the extreme, all people are the same insofar as they are members of the biological species *Homo sapiens*; while *these* people (i.e. in this group) are the same in that they all have blue eyes. Perhaps each individual is unique in being a unique *set* of generalisable properties, but as Wilkinson says, that is only contingently true. It is at least logically conceivable that another person exists, or has existed, or could exist (perhaps with the use of a tele-transporter device) with the same physical and psycholog-ical properties as I, and this is sufficient to undermine the claim in the argument against prostitution that each person is *necessarily* unique. Finally, Wilkinson also considers the role of a proper name in designat-ing a single individual. There is only one person, and has only ever been one person, who could be *this* individual, whom we happen to call 'Boadecia' (although there will of course be other individuals called Boadecia). But, as Wilkinson observes, this is an issue about language rather than about ethics; nothing of ethical import flows from the unique designation of this object.

Wilkinson's account is certainly correct *if* we make a crucial assump-tion about the world and about the purpose of philosophical enquiry. That assumption is that the world can be fully characterised in *objective* terms, that is, as a world full of objects, events, relations available for description and comparison by the philosopher. What is missing from Wilkinson's account is any acknowledgement of another person having a subjective point of view on the world. There is something 'that it is like' to be Boadecia, to use Thomas Nagel's striking phrase, and to look out upon the world from where she's standing; where she's standing not only in the room, but also where she's standing in her life, given how she got to this room with these people, given where she was aiming to get by being in this room here and now, given the determinate relationships and projects she finds herself in at this time. This is the root of the concept of uniqueness. It is a concept that is best grasped, not by surveying anonymous multitudes of strangers, but by considering the ordinary interaction with those we love and know well. I do not mean a generalised love of humanity, of course, but the sort of love that is conditioned by the acknowledgement of the other as having a radically unique point of view.

No, this does not meet the tele-transporter objection: it is conceivable that my beloved could be completely replicated. But to be honest, when other philosophers start resorting to fanciful science fiction examples to make their case, I am no longer confident about how I would think or respond to such examples to refute them. For me it is sufficient that such examples are not part of our ordinary experience of loving another and of acknowledging them as unique. In the same way, I do not accept Wilkinson's claim (p. 51) that 'I can think of no reason to believe that my lived experiences are radically qualitatively different from anyone else's'. If we look at Wilkinson's life from the outside, as a biographer or anthropologist, this is certainly true, and we are back to the problem of defining 'same' and 'different'. But surely Wilkinson's experiences *are* radically different from anyone else's insofar as they are and must be *his*. Someone else can have a very similar life as mine, a very similar personality, ambitions, values etc., but that does not at all threaten the fact that his life is his and my life is mine. The uniqueness of my life is presupposed in everything I think and everything I do. (Nagel's classic 1986 text *The View from Nowhere* is entirely concerned with this problem, and it is striking that Wilkinson does not even mention it.)

Responses to Warnock, Harris and Glover

Let us return to the HFEA clause designed to protect the child's welfare against the two known child abusers who have come to ask for IVF. The question becomes: *which* child's welfare? For no such child exists yet. There is a clear reason for the existing safeguards to prevent known child abusers applying for adoption, because there we can speak about a particular child who would be at risk. To put it another way, we have a child waiting for adoption, and we can compare its two possible futures: first, to be adopted by the child abusers, resulting in a good chance of suffering; second, not being adopted by the child abusers, and remaining in the (we may presume) minimally tolerable state institution waiting for a 'suitable' couple. The IVF child in question does not have two possible futures to compare, *but only one*: either he will be successfully brought into existence by this IVF procedure (assuming successful implantation and pregnancy), and will go on to face a real risk of suffering abuse, or he will have no future because he will not be brought into existence at all, either because the IVF is refused to the couple or because the IVF or the pregnancy fails. Indeed, it does not really make sense to say 'he will not be brought into existence' because there is no 'he' waiting off-stage for the privilege.

One way to speculate about the welfare of the child is to ask him how he's enjoying life – even if only hypothetically. Assuming the child is successfully brought into the world, and assuming he is indeed abused, what would he say if we asked him whether the IVF should or should not have been carried out? In other words, will he find his life sufficiently meaningful or pleasurable? Yes, he will wish that the abuse had been discovered earlier and that he had been removed from his parents, say. But that is not the event that we are asking him about; for that event is *in* his life, whereas we are interested in his views on the event that *began* his life. Will he wish that he had never been born at all?

He may. Some lives are indeed blighted in this extreme way. But I will boldly declare the number to be very, very small. I do not have any empirical evidence for this claim, and it is not clear how such evidence would be gathered. Perhaps one could start with the number of suicides, although even here it should certainly not be assumed that every suicide is a necessary result of believing that one's life has been blighted *from the start*: the usual reason will have to do with more recent events. (In Chapter 7, I will ask whether it even makes sense to speculate about the reasons, declared or underlying, for a suicide.) But for my purposes, I am seeking the reader's intuitive agreement that even child abuse victims can find life meaningful enough to go on, meaningful enough to have been given it in the first place. This is not yet to say that known child abusers should be encouraged to apply for IVF – merely to say that it does not make much sense to speak of the 'welfare of the child' in this context.

Now let me consider Harris's argument against the congenitally deaf parents who request the implantation of an embryo that has been tested to reveal the same predisposition to deafness. Let us assume that several fertilised embryos are in the Petri dish, some of which are 'deaf' and some 'hearing'. Implanting a deaf embryo would maximise the chances of producing a deaf child, if the implantation and pregnancy go well. Has the resulting child been harmed? In other words, is he worse off than he would otherwise be? The answer has to be no. In terms of possible futures, the deaf embryo will either develop into a child and then an adult or it will not develop at all because it has been discarded as part of the IVF process (or it has been implanted but naturally miscarries). There is no possible future in which *this* embryo can hear normally.

Harris compares this case to that of deliberately deafening a child, which he correctly takes to be harmful. But as before, Harris is proposing to deafen a child that already exists. That child is harmed because a deaf future is imposed on him in place of the hearing future that he could be reasonably expected to have otherwise. In the IVF case, Harris's mistake is to strive for an 'impartial view' of the situation, according to which a hearing embryo is always preferable to a deaf embryo, as if he was reading a shopping list of the features to be preferred in a bicycle: surely a bicycle with a seat is better than one without. But this is because bicycles are essentially replaceable: if my bicycle is stolen, you can buy me the *same* one in commiseration. 'Same' here does not mean the *very one* that was stolen (with its unique serial number), but the same *type* of bicycle, or at least a bicycle with most of the same generalisable features I was originally looking for in a bike (the difference between numeric and qualitative identity).

Harris speaks in terms of what would be of greatest benefit to the child, in this case, a pair of functioning ears. But it is not as if the child is first created, without ears, and then the Ear Salesman is invited over to demonstrate the latest models. Instead, it's an integrated package. But not only is Harris's metaphysics dubious, so is his focus of concern. But is this really for the sake of the resulting child, or rather for the sake of the parents, who are already planning their child's athletic or musical or intellectual future in competition with other children – and especially in competition with other parents?

The same logical mistake is made by Bonnie Steinbock (2002 p. 182) and John Robertson (1993 p. 164). Steinbock compares selecting against a defective embryo to advising pregnant women to take folic acid supplements during pregnancy in order to reduce the chance of the foetus developing neural tube defects. Robertson looks at the reverse side of the coin: in response to concerns that embryo selection will lead to efforts to select more intelligent, more attractive future people, Robertson reminds us that parents now have wide discretion to enhance offspring traits after birth with actions that range from the purely social and educative, such as special tutors and training camps, to the physico-medico as occurs with orthodontia, rhinoplasty, and exogenous growth hormone. Such actions [legitimately] fall within a parent's discretion in rearing offspring.

(1993 p. 164)

Both analogies fail. Steinbock's analogy fails because an embryo has a handicap right from the start – there is no 'room' for preventing *this* embryo from having this handicap, let alone curing the foetus of a handicap like Down's, although we might reach a point in the future where we could prevent the genetic mistake between egg and sperm that results in that handicap. Whereas folic acid is recommended to women already with foetuses. Again, Robertson's analogy fails because his improvements concern a specific child that already exists, and no other child is at stake; whereas selecting embryos on the basis of future intellectual or physical ability would be at the expense of those embryos not selected and therefore destroyed.

Choosing between two embryos is not like choosing between the two futures of a single embryo. Rather, it means choosing between the two respective futures of the two embryos. One future is of one adult who happens to hear, the other of another adult who happens to be deaf. It becomes very hard indeed to justify discrimination in favour of one adult and against another on the basis of the latter's deafness alone – unless adequate hearing is a requirement for a particular job, for example, which is hardly the case here. Indeed, it becomes impossible to justify such discrimination when the outcome for the loser will be annihilation. The two adults are ethically equal, and therefore the two embryos from which they grew are also ethically equal. And yet this goes against Harris's contention that it would be wrong to implant the deaf embryo when a healthy embryo was available. As such there is precious little ethically to allow a legitimate choice between the embryos, and so the parents' culture may well be enough to tip the balance.[6]

To be fair to Harris, he is partly aware of the clash of perspectives, and he tries to accommodate it by distinguishing between being *harmed* and being *wronged*. In deliberately selecting the deaf embryo, I am choosing a life that is harmed more than the hearing embryo. However, says Harris, I am not wronging the deaf embryo by selecting it. Wronging implies at least the theoretical possibility of redress or compensation, and here Harris acknowledges that it would make no sense for the deaf adult (into which the deaf embryo developed) to seek such compensation, since any option designed to avoid wrong would have prevented this adult from coming into existence in the first place; as such the putative wrong was, from the perspective of the deaf adult, inevitable, and therefore not wrong at all.[7] There is something to this distinction between harm and wrong: a woman is wronged when her husband cheats on her, even if she never discovers the fact and is therefore unharmed by it. But Harris presses it too far: everything that he says about harming can surely also be said about wronging. Harris, like other utilitarians, seems to think of harm (or pain, or suffering, or lack of welfare) as some metaphysical notion that does not in principle depend on a bearer of that harm; the individual is merely the contingent locus of the harm. As such one individual's quantity of harm can be measured against another's in absolute terms – and, within certain limits, this is the notion that I reject. The limits again refer to certain very rare cases of extreme defect or impairment where we can more confidently say that the individual's life is not worth living (and we can more confidently assume that the individual himself would agree). But allowing for the existence of such limits does not mean conceiving them as the thin end of a utilitarian wedge.

Robertson (1993 p. 76) draws a slightly different distinction between 'harm' and 'wrong', but ends up with the same problem. He cites an example from Derek Parfit, in which a woman is advised by her doctor to wait for a month before getting pregnant, by which time she will be off her teratogenic medication. In ignoring this advice, says Robertson, it is true that she has not harmed the child, since 'there was no way this particular child could have been born normal'. So far I agree. However, he continues, 'many would say that she has acted wrongly because she has gratuitously chosen to bring a suffering child into the world'. But first, it is not clear how the suffering that was wrongly brought into the world is different from the harm that was not; and second, if the alternative *for the child* was non-existence, then surely there is a good chance that he would consider the suffering (or the harm) a sufficient price to pay.

We now come to Jonathan Glover's claim that it might be a duty to abort a severely handicapped foetus if there were a good chance that the woman could become pregnant with a healthy foetus later on. Again, a lot of this discussion will turn on the word 'severely'. I have conceded that there is a very small number of foetuses afflicted with conditions sufficiently awful that their life can genuinely be said to be not worth living (although even this might not be clear-cut, if the mother wants to keep it). But Glover would include many more conditions in the 'severe' category than I would. So for the purposes of this discussion let's take a condition such as Down's Syndrome. On the one hand, it is legally 'severe' enough to justify abortion after 24 weeks' gestation in the UK, in accordance with the Abortion Act. On the other hand, if the Down's child does not have too many associated complications, he stands a good chance of living to 60 with a reasonable degree of independence, so it is simply implausible to say that their lives are not worth living.[8] So where does Glover get his strong moralistic tone from, saying that it would be *wrong* – not just unwise, say – to keep the Down's foetus?

A striking way to expose Glover's mistake would be to compare it to the case of *M*, described by Raimond Gaita (1998 p. 57). *M* was grieving for her recently dead child, and happened to be watching a television documentary on the Vietnam War. The documentary showed the grief of a Vietnamese woman after a bombing raid killed her children. And *M* declared 'But it's different for them. They can simply have more'. This remark could have slightly different meanings in different contexts, but in this one it was clearly racist. For *M*, Vietnamese children are essentially replaceable, like domestic animals; Vietnamese mothers do not notice the uniqueness of their children, they just notice the general features, and any child with similar features will do – indeed, the next child might even bear such desired features to a greater degree. The question is then whether Glover is advocating something similar: abort this handicapped child because you can always have another, better one.

Glover could reply that the situation with the Down's foetus is different precisely because it is still a foetus, while the Vietnamese children have already been born, started a life, enjoyed some of that life, and developed a relationship with their mother; in other words, foetuses are replaceable but children aren't. But as we saw in our discussion of McMahon in Part I, such an approach would seem to have *already* decided, by arbitrary fiat, that a foetus was 'below the threshold of respect' and therefore a replaceable non-person.

Two important points to note here. Unlike deafness, there is no suggestion that people with Down's form a separate culture that is worth cherishing and protecting; Down's is very much a defect and leads to various disabilities.[9] Second, I am certainly not arguing here that disability is a merely social construction, and that Down's people are merely variations on the norm. Unlike deafness, if there were a cure for people who already had Down's, there would not even be any suggestion of parents withholding it or of patients refusing it. And so if the health of the population is one's prime concern, then *any* measure aimed at reducing the incidence of Down's will be a good thing. But this is where Glover confuses the population perspective with the individual perspective, for he fails to notice the difference between removing the Down's from an individual who has it and removing the Down's by killing the individual. It is only the latter that Glover and the Abortion Act are encouraging. A doctor can assume correctly that no woman wants a Down's child (the indefinite article is the key); but he cannot assume that no woman wants *this* child, who happens to have Down's. In aborting the Down's foetus in favour of a later, healthy 'replacement', one is choosing one being over another, together with one being's future over another. As with my response to John Harris, there can then be no good reason to abort a Down's foetus in favour of a healthy one, because there is no good reason to kill the Down's *adult* (that it would probably become) in favour of a healthy adult.

Let me conclude this section by asking whether it makes sense to *regret* having had a child that turned out to have Down's. Amniocentesis can detect the presence of Down's in a foetus, but the procedure also carries a 1 per cent chance of inducing a miscarriage. For this reason many couples choose not to have it. Imagine that a Down's child is born to an unsuspecting couple, a couple who are not opposed to the abortion of defective foetuses. Does it then make sense for the parents to regret not having had the amnio?

The first problem with this is a more general problem about probabilities; or rather, about *living* with probabilities. If I smoke three packs of cigarettes a day for several decades, I can fully accept that it increases my chances of developing lung cancer by a lot, maybe more than 50 per cent. But then again, I might not get it. 'The statistics were about other people, not about me.' Indeed, I won't get it *until I get it*, so to speak. If I don't get it and die at a ripe old age from something else, then there will have been no point to reduce my smoking (at least, not on health grounds alone). And even if I do get lung cancer, there is always a certain amount of sheer bad luck that I can blame rather than myself, in the same way that one can blame bad luck for a whole host of other diseases and injuries.

The incidence of Down's is about 1 in 900. Is that a lot or a little? In most contexts that might be a pretty safe bet. If I could put down my life savings on a horse that had an 899 in 900 chance of winning, I probably would. If I were told that the budget airline I was boarding had a crash record of 1 in 900, I would pay more for a better safety record. Life with a Down's child is more difficult and frustrating, in all sorts of ways, than life with a normal child, but can it be equated with the loss of one's life savings or with the loss of one's life? I don't know what the answer is; different people will be risk-averse to different degrees, which is a fundamental objection to conceptions of morality or justice that depend on a standardised 'rational' response.[10] My main point, however, is that these sorts of calculations eventually become idle and irrelevant once the child is born – with or without the Down's. With the bet on the horse, it makes sense to reconsider the matter in hindsight, to re-examine options, to regret the choice I made. Imagine my gangster uncle has connections with the bookmakers who took my bet and has a quiet word with them, and they return the money: this would mark a full return to the *status quo ante*. With the arrival of the child, on the other hand, there is no going back. And therefore there is no 'room' for regret; even if you consider the amnio decision to have been a mistake, there is nothing to learn from the mistake. Even if one learns that having the amnio is wise *next* time, this does not affect the situation with the first child. If I had done the 'wise' thing back then, *this* child would not be here today. And nothing could have been done back then to cure this child, so from the perspective of the child's short life, there was no room for the 'wrong' decision and therefore no room for regret.

Besides, not only is it appalling to wish that someone had never been born, but the parents are especially debarred from formulating the thought, since they have to live with the child for the rest of their lives. This returns me to the discussion of 'learning to love' in the last chapter. By giving the parents the choice, prenatal screening increases intolerance of imperfection and encourages parents to regret what the child is not rather than to love the child for what it is.

The paradox of non-directive counselling

In the last scenario the assumption was that *if* the parents had elected for the amnio, and if this had confirmed the Down's in the foetus, the parents would have gone ahead with the abortion. In this section I want to look at the problems associated with these sorts of tests, and what it might mean to counsel parents about the decisions that can be made on the basis of the test results. The point of such counselling, the British Society for Human Genetics says, is to allow a couple to 'talk about what the result means, and what options are available'.[11] (For my purposes it is immaterial whether the condition is a hereditary one or not; I will not be exploring the further ethical question of whether the parent should inform his or her siblings, for example.) The counselling offered by the service is supposed to be 'non-directive'. This means

> you will not be 'directed' or told what decision you should make. Genetic professionals are *not* in the business of trying to persuade people. Our role is to try to explain the facts as clearly as possible, giving the person or family accurate information on their options in a way which they can understand, and helping them to make up their own minds.
>
> (Ibid., italics in the original.)

Clearly one purpose of genetic counselling is to provide early warning of the birth of a child with such a condition, and to discuss what it means to live with that condition, so that the parents have time to prepare themselves psychologically and logistically, as well as to inform friends and family. If the parents are not sure what to do, however, the counsellor is supposed to withdraw after presenting the facts and options, and this alleged non-directiveness is what I would like to explore in this section. I will suggest that the ambition to provide 'just the facts, ma'am' is disingenuous, as is the effort to 'help them to make up their own minds' without 'persuading' them.

Let me recapitulate some of the discussion of Chapter 1 on ethical expertise. In most areas of life, there is nothing in principle as too much information. Yes, a lot of it may be irrelevant or dull, but it's always nice to have the option to ignore it for being irrelevant and dull. More information, especially about available options and the likely consequences associated with each, normally means being more informed, which means that I can make decisions and consent to proposals with greater genuine autonomy. When buying a car or a house, there is no limit in principle to the amount of information that would benefit the decision. The only limits will be contingent: those involving an ability to understand technical information, and those of trying to keep all the relevant information before one's consciousness at one time while making the decision. And even when the technical information cannot be fully understood, as in the case with most treatment decisions in medicine, a skilled communicator can at least re-package the options by stressing the likely impact of each on the patient's experience: option A is more painful, but is likely to be more effective, than option B.

In terms of non-directivity, this too would seem to be an uncontroversial concept. Certainly a mortgage advisor could give the relevant information without offering advice. A survey has revealed that the lovely house at 221b Baker street has a dodgy roof, and therefore you might want to reflect about what that will cost in the future. However, whether or not you buy the house, with the dodgy roof and the associated future costs, can still be as personal a decision as the choice of décor, and the mortgage advisor can only re-describe the options in more detail if necessary. It's your money. And if the house is attractive for many other reasons then the roof just becomes part of the price you are willing to pay for it. Analogously, the genetic counsellor can arrange for a 'survey' of the foetus, can report on a dodgy chromosome 23, and can describe how much more that will cost in the future, in terms of money and time and effort, than a healthy child. And after presenting the information, the counsellor discretely withdraws without any attempt to offer further guidance on a purely personal matter for the parents.

However, there are two crucial differences between the dodgy chromosome 23 and the dodgy roof, and those are, first, the fact that the latter can be *repaired*, even if we cannot afford it right now. There is a clear procedure and a clear cost of the repair. Even if there is always a risk of botching the repair job and having to repeat it, eventually one can expect the roof to serve its function. Second, if the roof gives us too much grief, we can cut our losses and sell the damn house and buy another one. Neither option is possible with the dodgy chromosome 23 and its hapless bearer. The only way to get rid of the Down's is to get rid of the individual with it.[12]

By why should this fact about the non-curability of Down's undermine the alleged non-directedness of the counselling? Different people will take the information in different ways and will make different decisions, depending on their ethical beliefs, *a basta*. Glover and I would both reject such a model, but for different reasons. Clearly Glover would favour some robust counselling ('moral persuasion', as he calls it [p. 146]) in the event of positive test results, but he stops short of calling for compulsory abortion because of the likely disutility of such a policy on the wide scale. Interestingly, he imagines 'horrors' associated with any attempt to implement it, partly because it would go against the will of some of the pregnant women and 'perhaps against their moral principles'. So although he was quite confident in the ethical advice he offered his readers – and by implication, the women pregnant with a defective foetus – he now suddenly decides to take seriously their moral principles even if they evidently contradict his. But it is a *grudging* concession rather than a serious one; the implication is that these principles cannot be shifted by reason alone and are therefore closer to inherited irrational prejudices than to genuine, fully endorsed moral principles. They have to be taken seriously, but only as trees and fast-moving vehicles have to be taken seriously when deciding which route to drive. My approach throughout this book has been to take other people's deeply held ethical beliefs seriously, and to start examining such beliefs from within their perspective – rather than as contingent obstacles to rational progress.

Now we are getting to the paradox at the heart of non-directive counselling. For a woman who has strong ethical beliefs about the impermissibility of abortion, the whole practice of putting 'just the facts' on the table, of allowing the client to 'review the options available' is already directive: for it suggests that abortion is an option that is at least worth taking seriously, even if it is to be subsequently rejected. It suggests that the option to destroy a foetus *solely* on the grounds of its handicap is an ethically legitimate option that is worth contemplating, even if it is eventually rejected. For each option there will be reasons in support of it and reasons against it, and each responsible moral agent should not be afraid to discuss them with another, e.g. with a genetic counsellor. We have already seen the difficulties in striving for a neutral language (in Chapter 3); now we are considering the attempt to provide a neutral list of options. Here is Peter Winch:

> If we wish to understand the moral character of a particular man and his act, it is, often at any rate, not enough to notice that, for such and such reasons, he chooses a given course of action from among those he considers as alternatives. It may be at least as important to notice *what he considers the alternatives to be* and, what is closely connected, what are the reasons he considers it relevant to deploy in deciding between them.
>
> (1972 p. 178, italics in the original.)

In many ethical disagreements, the two disputants may well agree on the options available and disagree about which would be best or right to perform. But this is not one of them, for the simple reason that many people do not consider abortion to be an option; or they consider it an option in certain, very serious circumstances such as a threat to the woman's life – but that is not what the counselling sessions are for. Or they consider abortion permissible in principle but find they just cannot go through with it in this case (perhaps they do not feel comfortable with being given such legally-sanctioned power over another's life). Winch's point about the reasons reminds us of the conflict (in Chapter 2) between the manager and the doctor over the latter's absence from work to tend his ill son: a reason is good enough for one, but not for the other, and it is hard to find fault with either of them.

For the NHS to even offer prenatal testing, under the guise of ascertaining and conveying the 'facts', makes it sound innocuous, just another routine investigation. But the implication is that anybody who even refuses the testing is somehow being irrational; or worse, they are being ethically irresponsible, for they do not want the best for their baby – in the same way as a woman who smokes heavily during the pregnancy is being irresponsible. Then, once the child is born with its genetic condition, the implication is that such a condition could have been prevented, and the kind, reasonable counsellors would have made everything so much easier. One point of the testing is preparatory; but beyond that, there would not be any point to the testing unless it was to send a signal about an option being ethically worth considering and therefore in some cases to be implicitly recommended. Again we have the clash of perspectives. Refusing to have one's foetus tested for a genetic disorder, and refusing to abort one's foetus after a positive result, is not the same as irresponsibly damaging an otherwise healthy foetus; only in the latter case could anything have been done to improve the life of *this* foetus (e.g. by stopping smoking).

Interestingly, the pressure to accept the implicit recommendation of the allegedly non-directive counsellors is even stronger in those countries without a national healthcare system. For a private healthcare insurance contract might require the parents to have the test, and make them partly liable for the ensuing care costs of any disabled child that resulted – as if the parents had *chosen* to keep the child in the same way as they had chosen to go bungee jumping. Needless to say, the parents in such a case would probably not have conceived this as a choice.

With the implicit recommendation comes an implicit value judgement about those individuals with the handicaps in question: a life with Down's is simply not worth living, the opinion of the Down's child is not worth listening to and so there is nothing wrong with cutting that life short. Sometimes the judgement is couched in terms of the chronic suffering of the Down's children and adults, a suffering that one can be justified in preventing. But it is simply not true that all individuals with Down's suffer in the brute physical sense implied, nor is it true that those who do suffer cannot be treated with routine pain relief. How then to avoid the implication of allegedly non-directive screening that abortion is to be justified with reference to a judgement about the ethical disvalue of *anyone* with Down's, for all of them were once foetuses with Down's (Wyatt 2001). Although as adults they are protected from being killed off when their existence is inconvenient or expensive or disappointing for others, that protection is arbitrarily time-indexed.

There is a larger paradox here about disability. Even if most of us can fully respect a disabled adult, can treat him as a moral equal, can support and indeed fight for the rights of the disabled, most of us would can also say (at least *sotto voce*) 'I'm glad I do not have your condition' or 'I hope my child is not born with your condition'. Some handicaps such as deafness can be relatively mild and can allow membership in a distinct culture, but that is surely the exception. Most handicaps, from a population perspective, are unambiguous bads. But all handicapped people, from their individual perspective, cannot consider their handicap (if incurable) to be bad in the same way: for them it just is a fact of life.

Resource allocation and the clash of perspectives

In this final section, I want to consider a huge topic in medical ethics, that of resource allocation. I do not want to get embroiled in it, however, because I do not have much to add to the existing debates which are well covered elsewhere. What I do have to say will not have anything to do with embryos or foetuses, but will concern the clash between the population and the individual perspective. However, it should not be expected that my argument will generate, let alone justify, specific recommendations for *policy*, and perhaps this is an inherent weakness of the whole approach I adopt in this book. However, I do like to think that some of my arguments could provide the groundwork for further applications in policy questions.

The problem as described in most textbooks is quite clear: public healthcare resources – money, space, equipment, staff, etc. – are scarce. What is the best way to allocate them? The usual answer comprises a principle of greatest cost-effectiveness (i.e. greatest benefit given the money available), as limited by a principle of fairness, where fairness involves the minimal sense of aiming to treat like patients in a like manner. The likely benefits to patients, can, at least in theory, be measured and compared, and a combination of treatments can be chosen that will generate the greatest aggregate benefit. How this works in practice is very complicated, I imagine, and I won't get into it; but let me assume that an ideal allocation can be found and successfully justified. Obviously we are dealing with the allocation of resources across an entire population, and so we have to take a population perspective.

Now imagine the following situation. A patient is diagnosed with bowel cancer. He has read on the Internet of an expensive new drug, Z, which has a 1 per cent success rate in certain types of bowel cancer. In the UK, the cost-effectiveness of drugs is determined by the National Institute for

Health and Clinical Excellence (NICE), who regularly revise the guidelines they issue in the light of new evidence. In this case, the guidelines are not in favour of prescribing Z to this patient, given the type of cancer he has (and let us assume they do not favour any other drug that might have a decent chance of success). The doctor explains the situation to the patient. But the patient is adamant: 'I don't care about fairness, I don't care about cost-effectiveness, I want everything to be done for me! One percent is a good enough probability for me!'

Now I suggest that it is possible for this patient to accept the expertise of the medical profession, that is, to accept the principle of the NHS having to make tough decisions about allocating the scarce resources. It is also possible for this patient to understand and accept the basic principles of pharmacology, the basic principles of measuring and comparing cost-effectiveness, the clinical trial data used by NICE, the actual NICE guidelines and the budgetary limits of the NHS. He can accept all this at an intellectual level, when considering the matter from the population perspective. But when he finds himself in the life-threatening situation, then the population view no longer commands such authority, for accepting the population view will literally kill him.

As the doctor treating this patient, I could respond 'tough luck' (although perhaps not using exactly those words). Indeed, I could get annoyed: 'what makes you think that you deserve special treatment, at the expense of others?' After all, one of the requirements of justice is that the *loser* should also accept the just allocation decision, however grudgingly. If I give you a piece of the cake, then I can deny giving you a second piece no matter how much you want it so long as there are others at the table who have not received any cake yet – and I can explain the reason for your not getting another piece, and legitimately expect you to accept that reason. But when the issue is not about how much cake someone gets but about whether he will live or die, it becomes absurd to assume that he should accept the population perspective without demur. To put the point slightly more technically, it is absurd to think that the reasons generated by the population perspective will necessarily find 'purchase' in the individual.[13] Yes, it would be *admirable* for him to forego the treatment in favour of the little girl in the next bed who needs it more than he does; but when the crunch comes can he really be *expected* to give his life for her, as impartial population ethics would seem to enjoin? If he refuses, I can call him selfish, but if it is his life at stake, such an accusation will just glance off him: of course he's selfish, but what does that accusation achieve? To find purchase, the accusation of selfishness must presuppose a future in which the accused can be expected to be less selfish; but by refusing to be selfish he loses any hope of such a future. And while fatal self-sacrifice can make sense in many contexts, the object of such self-sacrifice must surely be something intelligible and appropriate, such as one's kith and kin (if the other claimant is my sister, say), or one's God or one's nation. But it is not clear how one can sacrifice oneself purely for justice; very few of us are saints.

What I am saying is not controversial if it is merely a psychological description of what people *do* do – put themselves first when they are desperate. It becomes controversial when I defend the individual perspective as having *equal authority* as the population perspective in life-and-death situations. By 'equal authority' I mean that the loser – in this case the man with bowel cancer – cannot be reasonably expected to accept the population perspective as having priority on *his* life, since his life is the only one he has, and it is under direct threat. If he dies, then he loses any capacity to respond to any ethical arguments in the future. Therefore his resistance to the claims of impartial justice is not merely a matter of self-interest, as in the cake example; instead, because it concerns life and death, it becomes an ethical matter. Above all, he *knows* that drug Z is out there, just as he knows that the NHS does have the money, just as he knows that the NHS will not crumble if they make an exception on this one occasion. Yes, of course the NHS would collapse if exceptions were made for everyone, but the NHS is not going to, is it – so it can afford to save him.

There is an interesting comparison here with the attitude of Western policymakers to refugees from war-torn parts of the world. If one considers the individual refugee alone, it is hard to find any good reason *not* to let him in, where I am taking a reason as one that would persuade the refugee himself: the UK government can easily afford to clothe, house and educate him without any real threat to the well-being of existing UK citizens, and he will be better off here, politically and economically, than in his country of origin. But from a population viewpoint, exactly the same could be said about *millions* of potential refugees in all parts of the globe. The principle of fairness would require the UK to let in all like cases, or no one at all. And so the UK's refugee policy is a fudge, typical of most Western countries: if a person is able to get to UK soil, by fair means or foul, and apply for refugee status here, then he will be duly considered for it. But anybody who can't make it this far will not be considered.

There is a final irony to point out in the topic of resource allocation. As a topic in the academic field of medical ethics, it is certainly one of the largest. There is a chapter on it in every medical ethics textbook and it is taught to UK medical students in every medical ethics course. This makes sense: medical students should understand something of the routine problems encountered by NHS management and will themselves have to make more and more of the same sort of decisions as they progress in their career. It is certainly better to organise such thinking using consistent principles rather than gut instincts.

And yet here is the irony. An individual doctor is *not supposed to* make a treatment decision on the basis of a scarcity in resources. The only basis for non-treatment, apart from the patient's refusal, is supposed to be clinical contra-indication or futility (together with the patient's best interests). This fact is reflected in the advice that doctors are given by medical lawyers representing the NHS Trust: it is much safer, legally speaking, to justify a non-treatment decision on the basis of futility since this is closer to the terrain of the doctor's unquestionable expertise. Whereas an individual doctor has neither the authority nor the information nor perhaps even the expertise to implement a 'personal' resource allocation policy on the hoof, inconsistent with his colleagues' policies and with the wider Trust policies.

In other words, we are teaching our medical students about resource allocation, but then later we tell them to assume that resources are infinite when deciding whether to treat *this* patient or not. They are actively discouraged from saving public money. It also leads to a curious kind of double-think, and to inevitable conflicts between policymakers and managers on the one hand, and clinicians doing their best for the individual patient on the other – the good cops and the bad cops. The resolution of such conflicts in practice would then seem to leave a large space for the contingent personalities of the two cops involved. After all, any clinical judgement that Drug Z is futile will be a matter of setting an arbitrary standard, that such-and-such per cent is not sufficiently likely to produce sufficiently good results, given the cost of Drug Z (assuming that Drug Z is not *completely* futile). But if the patient is desperate enough, then he may think such a chance worth taking, whatever the cost.

5
The Abortion Debates

We have examined the subject of abortion indirectly in several other chapters, but it is now time to step boldly into the hornet's nest. This indirect approach was necessary, I have claimed, because the meaning of the act of abortion itself – even before we get to the rightness or wrongness of it – is so complicated that one first has to build up a more nuanced understanding of pregnancy, birth, childhood and parenthood. Although this understanding is still insufficient for the full treatment that I would like, it should be enough to show how simplistic and wrong-headed some of the dominant mainstream arguments about abortion can be, especially in their insistence on pigeonholing all disputants into opposing camps of 'pro-life' and 'pro-choice'. The ultimate impotence of the mainstream arguments is obvious in the debate's intractability: there are always well informed and reasonable people 'on the other side' whom my arguments cannot reach, and this fact alone should suggest that a different take on the problem is in order. In what follows I shall not make any effort to present a systematic and comprehensive survey of the arguments, since this book is not a textbook; but I will be considering some of the main arguments before commenting on the tacit assumptions made by their proponents.

In parallel to other issues covered so far, there are at least three abortion debates: one surrounding the question of whether abortion in a certain type of circumstance and/or for a certain type of reason is ethically permissible or not, a second about the formulation of the best public policy that will satisfy the greatest number of people and offend the least, and a third, surrounding what a particular woman or couple ought to do in particular circumstances. Too often these debates are elided, especially by philosophers who claim that ethics should be essentially impersonal, a conception of ethics I have challenged in Part I. It is the personal debate that I am primarily interested in, in keeping with the book's emphasis on the experience of ordinary people, and their description of their own experience in ordinary language. In addition, because of the importance of ordinary language, I will continue to use the word 'abortion' rather than the more clinical 'termination'. If clinicians prefer to avoid the negative resonances and connotations of 'abortion', it is precisely these resonances and connotations that interest me. The paradigmatic situation, therefore, is that of a woman asking a close female friend or family member for advice about what she should do, *given* a liberal abortion law such as that in England. In other words, she already has the right to an abortion, and now what should she do? I thus want to side-step the debate about whether or not this legal right should be granted to all in the first place. This priority also allows me to avoid the philosophically problematic notion of legal rights, precisely because she is unlikely to use the concept of a legal right in the conversation with her friend.

My discussion will be taking place against the background of the English law, and I shall assume the law to be roughly similar to that in many other Western countries. So it is worth spelling out the relevant details. The Abortion Act 1967 (as amended in 1990) permits abortion before the end of the 24th week of gestation, in order to avoid 'injury to the physical and mental health of the pregnant woman', where such injury is to be judged as sufficiently severe and sufficiently likely by 'two registered medical practitioners'. The 24-week threshold was chosen because it corresponds roughly with the threshold of 'viability': that is, from the 24th week a foetus has a good chance of surviving premature birth with manageable conditions, so the logic presumably is that a pregnant woman who does not want the child after that date can, in principle (though not in practice), have the child induced and then removed for adoption. In the UK the vast majority of abortions are carried out prior to the 24th week, precisely on the grounds of preventing such 'injury'. In practice, 'injury' is interpreted extremely broadly and allows for virtual abortion on demand. The NHS pays the entire cost of the procedure. After the 24th week, abortions are only permitted (i) to prevent '*grave permanent* injury' to the woman (i.e. not just 'injury'), or to save her life, or (ii) when there is a 'substantial risk' of the resulting child being born 'seriously handicapped'. There is clearly some room for disagreement about how substantial the risk has to be and how serious the handicap. A genetic condition will result in a 100 per cent certainty, but this leaves the question of seriousness. In practice, a condition like Down's is considered sufficiently serious to legally justify abortion.

What is striking about abortion in the UK, unlike in a country like the US, is that it really is not a big political issue. By this I mean that, as far as I know, there has been no serious political move to significantly alter, let alone repeal, the Abortion Act in the 40 years since it came into force. (The only change was to move the main cut-off forward from 28 weeks to 24 weeks in 1990, to reflect the change in the viability threshold as a result of improved neonatal technology.) There have been no major political demonstrations in the streets to tighten or loosen the law, and there have been very few academics on BBC Radio 4 suggesting major changes. While there is general concern at the sheer numbers of abortions, and while the original drafters of the 1967 Act did not anticipate the present situation of virtual abortion-on-demand, the political responses have usually had to do with improving sex education or making contraception more available, both with mixed success.

Alongside the serious risk to the mother's health and the serious risk of foetal defect, there is a third category of pregnancy which generates strong reasons for abortion, even after the 24th week, and that is when the child is a result of rape (where this includes sex to which a girl cannot be said to legally consent because she is underage). Although not explicitly mentioned in the law, one can imagine the resulting birth causing 'grave' and 'permanent' psychological injury to the woman. In what follows, I will take these three types of cases to be relatively unproblematic and will concentrate on abortions requested for other reasons, usually to do with the great inconvenience and cost and responsibility of having a child (or *another* child) in one's life.

Arbitrariness and potential

I want to pick up a discussion from Chapters 1 and 3, where I introduced the problem of defining personhood or defining the criteria for moral status. Imagine a line to represent the development of the human being *in utero*, from the separate sex cells at one end to the newborn infant at the other. For the purposes of this discussion I shall assume that the separate sex cells have no special moral status and (*contra* McMahon) that the newborn infant has full moral status. 'Moral status' is a clumsy expression for something that should be obvious: newborn children should never be deliberately killed and should only be allowed to die under very rare circumstances where they are suffering terribly and their prognosis is very poor. Once framed this way, it is tempting to ask when the foetus acquires this status, and with it the protection against deliberate killing.

The law draws the threshold roughly at the foetus's viability. Another key legal threshold in the UK is at 14 days, corresponding to the appearance of the 'primitive streak', the beginnings of the spinal cord; beyond this threshold the Human Fertilisation and Embryology Authority does not permit experimentation, e.g. in stem cell research. A third threshold might be the 'quickening' between 16 and 20 weeks, the first movements that the woman can feel (although muscles start contracting much earlier), and traditionally representative of the moment when the soul enters the body. Jennifer Jackson proposes the eighth week, at which point all the major organs are present, and the foetus 'looks like a baby, inside and out' (Jackson 2006 p. 128). Other thresholds might be the first beat of the heart, or the first electrical activity of the brain – corresponding to the two most famous ways in which the *death* of an individual is determined. Finally, Mary Warren (1973) is famous for proposing a threshold of birth itself, since before that age the foetus remains physiologically dependent on the woman and therefore part of the woman. Indeed, like McMahon, she sees nothing intrinsically special about infants either. Both foetuses and infants are 'outside the moral community' (Warren) and 'below the threshold of respect' (McMahon). The only thing about infants, concedes Warren, is that giving them up for adoption is straightforward and always preferable to killing them when they are unwanted.

What on earth is one supposed to make of all this? Is there any one of these thresholds that clearly wins out? Or is the very plethora of possible thresholds an indication of the essential *arbitrariness* of trying to determine one? Certainly there are good scientific reasons for trying to organise foetal development into stages, but why on earth should one think that any *ethical* significance has to be attached to those stages?

The only non-arbitrary point is conception. Prior to that there was no single individual that would necessarily continue developing if the conditions were right. In speaking of an 'individual' here, I am using not a moral concept but a biological one; there is a clear sense of a single organism developing through stages. But this biological point has important consequences for the way each of us see ourselves – and this is an ethical significance. It is uncontroversial that we can speak of our childhoods, even if we cannot remember the events attributed to us: 'when I was very young, apparently, I could never pronounce "th" properly'. In the same way, there can be meaningful discussion of what we were like *in utero*. A mother will reminisce and tell her adult child: 'I remember when you first started kicking, your father and I were so excited'.

One way to oppose the search for a threshold in trying to clarify the abortion question is to speak of *potential*, and the most famous argument from potential is that of Don Marquis (1989). The point I made in the previous paragraph seems to be the beginning of an argument from potential, but Marquis's is importantly different. Marquis suggested that the wrongness of killing an adult was that it deprives that adult of the 'future goods of conscious life'. And therefore killing a foetus does exactly the same thing. This has not been a very successful argument, mainly because it is vulnerable to three rebuttals. The first is to say, quite simply, that an acorn is not an oak tree or that a cake mix going into the oven is not the same as the cake coming out. The destruction of a dozen acorns is hardly as serious as the destruction of a dozen oaks. If I spill the cake mix onto a dirty floor, it is not the cake that is lost. The second rebuttal is to say that, although I now have the potential to be Prime Minister, and to enjoy the goods of my conscious life as Prime Minister, that is no reason for me to have all the rights and responsibilities of being Prime Minister *now*; I first have to become Prime Minister in the usual ways. Third, it is odd to say that the wrongness of killing an adult has to do with that adult's future: does the wrongness not have anything to do with *him*, i.e. everything he is, his past and present as well? After all, murdering someone who is terminally ill is still murder, even if they do not have much of a future to lose.

These three rebuttals are well-taken against the argument, at least as phrased by Marquis. However, these rebuttals neglect the two crucial facts about human lives that I discussed earlier, and these facts are important for my own version of the argument from potential. First, individuals have a perspective on the world and on themselves that is sometimes at odds with the population perspective presupposed in most debates of medical ethics. So from the population perspective, of course a potential X is not the same as an actual X. But when an adult considers how his mother almost miscarried due to a serious illness during pregnancy, it is the potential adult who was threatened then and who now feels a distinct uncanny feeling. Second, human beings live a life that is essentially diachronic, that is, embedded in time passing, and this fact affects the *meaning* of the stages of life. This second point ties in with my discussion of the miraculous in Chapter 3. Infant human beings may not be aware of their future, but everyone else is. Right from its entry into the world, most infants are the centre of attention, as well as discussion and speculation about what the infant will *become*, for better or for worse. (And so much of the tedium and difficulty of looking after an infant is tempered by the knowledge that this stage will not last forever!) So the meaning of the infant stage contains a direct reference to its future, even if that future does not exist yet. And exactly the same may be said for the foetus, for it is the same biological organism. The excitement at the 20-week scan is an excitement about the future, not about the present. This, then, is the core intuition of my argument from potential.

Both of these points can be nicely elaborated by considering Jonathan Glover's ironic suggestion (quoted in the last chapter) that he himself, as an embryo, ought to have been deselected and destroyed if his parents had had a choice of embryos and reliable ways of testing them in order to find one that was more talented. Judging by his tone, Glover thinks that this is an accidental oddity, but I suggest that it marks the ground for a serious *reductio* against his position.

Glover's comment is clearly from the population perspective and he is neglecting the very life in which he himself has been living all these years. From an individual perspective, it is too late for Glover to take seriously the possibility of his non-existence. His existence is not only a *necessary*, but a *necessarily valuable* feature of the world. Of course his existence is highly contingent – his parents could easily not have met, the particular sperm might not have won the race, the fertilised embryo might not have implanted, he might not have been rescued from his crib during their house fire etc. – but these possible worlds never existed. Of course he was just one embryo among several hundred thousand born on the same day, and of course he was just a 'bundle of cells' of no importance whatsoever, either in comparison to the rest of humanity or to the size of the universe. And yet from his perspective, now, *in medias res*, his existence, right from the start, has to be not only hugely important, but necessary and necessarily valuable; for that double-fact is presupposed in every thought he has about the world, including every thought that he has about his non-existence. Even the possibility of deselection is ethically outrageous, since, when looking back on the embryo, it is as if his future already existed at the time, even if, from the embryo's perspective the future is unknown and will only unfurl one day at a time. To summarise: there are no reasons for destroying the Glover embryo that will be good enough for the adult Glover.

This provides a further element to the argument from potential. Unlike Marquis, I am not claiming that the embryo has moral status from conception because it has the mere potential to become an adult human being – that claim falls into the obvious problems mentioned before. Rather, I am claiming that the embryo has moral status from conception because there is a *particular* adult human being, into whom it would probably develop if allowed to, who would be capable of justified protest against its destruction. Just because that adult is in the future, and just because we may never hear his voice if the embryo is destroyed, does not mean that he can be ignored.

If this loose conception of time sounds odd, it might help to consider another context where we make plans for the future without knowing anything about the beneficiaries of such plans since they do not exist yet. Imagine I build a house for my family, including my two preschool daughters. I can meaningfully hope that the house will also serve my future grandchildren, when they come into the world; and if the house burns down I can meaningfully regret the loss that these future beings will suffer.

Women and mothers

Alongside the set of arguments centred on the search for the best threshold is another set focussed more on the pregnant women. Ironically, many participants on either side of the abortion debates seem to ignore the fact that foetuses develop in women – and this was important for my understanding of the foetus and child as 'of the flesh' of the mother. The debate would be the same, they might argue, if foetuses developed in jars on shelves. But of course the woman is directly involved in the whole business, and this leads to a number of distinctive arguments, all based on accepting that the foetus does have moral status, but that it can nevertheless be permissible to kill it. (Sometimes these arguments are called 'feminist', but this term is itself so vague that it does not help to clarify much.) Once again, I will limit myself to a brief account of some of the main arguments, for they have already been presented in much greater detail by others.

(i) *The backstreet abortion argument.* One reason often advanced for legalising abortion is to eliminate dangerous 'back street' amateur practices and thereby protect women. The argument runs: desperate women are going to seek abortions anyway; therefore we might as well legalise it, regulate it and make it safe. And indeed from certain statistical points of view, it is now safer than natural childbirth. On its own, it is not clear whether this is a strong argument. After all, heroin use is dangerous, people are going to use it anyway, but those facts in themselves are not a reason to legalise heroin.

(ii) *The argument from freedom and control.* Women ought to be allowed more control over their reproductive systems, not only in terms of the ease of using contraception, but of the ease of getting rid of the results of failed contraception. Only then can a woman enjoy the same promiscuous freedom as men and can choose to have children if and when she wants to.

(iii) *The socio-economic argument.* On the whole, life in Western societies is more difficult for women and mothers than it is for men, and this is not fair. On the one hand, women are still expected to do the bulk of the child-rearing, free of charge; on the other hand, they are given very little and very poorly remunerated maternity leave; childcare is expensive; shops and workplaces are not designed for breastfeeding or pushchairs etc. Having and keeping a child then becomes a massive disruption to most women's lives. Given the difficulties in building an interesting career and achieving promotion for *childless* women already, or in just getting by as a single woman on a low income, it is no surprise that some women choose not to have children (or not to have any *more* children). If the state took better care of mothers and children, then many pregnant women – not all, of course – would not need to consider abortion as an option.

(iv) *The Good Samaritan argument.* In the famous example from Judith Thomson (1971), a woman goes into hospital for an appendectomy. After she is anaesthetised, there is a mix-up in her medical notes, and she is taken for a 'body donor'. When she wakes, she finds herself hooked up through many tubes and machines to a stranger lying unconscious in the neighbouring bed. This stranger turns out not only to have full moral status, of course, but is also a famous violinist and thus a useful member of society. The hospital have since discovered the mistake, and come to explain the situation. They're terribly sorry, but now that she is hooked up, would she mind seeing it through? After all, it will only take nine months; and if she leaves now, the violinist will die. Thomson's argument is that, while it would certainly be admirable for the woman to play the Good Samaritan, there can surely be no *duty* for her to rescue a stranger at such cost to herself.

The main problem with arguments (ii) and (iii) is that they do not distinguish sufficiently between the foetus and the newborn infant. I am still assuming that, *contra* McMahon and Warren, most of us would consider an infant to be a full member of the moral community; in fact, in

one respect it is worth more than many adults in virtue of its pure innocence and unfulfilled potential. If a woman wants to lead a life of wild
promiscuity or career devotion, that would not in itself be a sufficient
reason to kill a newborn child that would impede such plans. If a
woman believes that she cannot afford to house and feed a newborn
child, then that is not a reason to kill it. In both cases we would be
tempted to respond 'too bad, whether you like it or not you are now a
mother' – or else we would advise giving the child up for adoption.

The main problem with Thomson's violinist example is that a foetus
is hardly a *stranger*; as I discussed in Chapter 3, there is perhaps no more
intimate relationship in human society – and with that must surely
come some notion of obligation. In one sense the baby that is born to
the woman *is* a stranger, for the woman has never seen it before; but
this would hardly be grounds for denying all duty to rescue it.

In order to forestall accusations of conservatism and sexism, let me
stress right away that I am entirely in agreement with the spirit of argument (iii), although I shall not argue for it at greater length here. It is a
lingering scandal that the UK government has been unable to enforce
equal pay legislation and that it is so miserly in its support of mothers
and young children. And although fathers are probably taking a much
more active role in child-rearing today, the broad expectation that it
remains primarily women's work continues.

Finally, the problem with all three arguments is that they do not focus
sufficiently on the *attitudes* to the potential life in question. In the next
section I want to develop this point of Rosalind Hursthouse (1991).
Hursthouse rightly claims that in some cases of abortion, the issue is not
about whether the woman has the *right* to an abortion or even about
whether an abortion would be 'for the best', but rather, about what we
are to think of such a woman – for example, that she may be selfish or
callous. I am less confident than Hursthouse that such ascriptions can
form the basis of a virtue-ethical theoretical approach to the question,
but it certainly becomes philosophically revealing to consider what this
callousness might consist of and where it might come from.

Attitudes to life

The outlines of a different abortion debate are slowly emerging. Not a
debate about the act and its ethical permissibility, but about whether
the woman and the doctors in question truly *understand* what they are
doing in asking for or performing or approving abortion. This is not
at all to suggest that *if* they understood they would not do it; not at

all, abortion may still be necessary, but it would be done with a much heavier heart. There is then the paradoxical of whether it is a bad thing or not that some women and some doctors do not actually understand what they are doing.

The problem, as before, is that any 'debate' is probably not about an exchange of arguments, but about different *ways of seeing* – exactly as the Vegetarian and the Carnivore saw the same chicken in different ways, without the implication that one of them had a privileged or correct view. The question in that 'debate' was whether the Vegetarian could bring the Carnivore to understand what he was doing when he was eating meat, where such bringing-to-see *might* be achieved by inviting the Carnivore to a slaughterhouse. The 'might' is important, for there can be no guarantee that the Carnivore will become a vegetarian after such a visit; nor is there necessarily any failure of cognition or rationality if he fails to do so. However, chances are he is more likely to understand what his meat-eating costs in the wider sense and he is more likely to understand *what* exactly he is eating. This returns to my distinction between the different kinds of knowledge. Nobody would deny that butcher's meat comes from animals, but few people have direct experiential knowledge of the transformation from animal into meat.

So let me consider a similar scenario. A woman at 20 weeks' gestation comes in to ask for an abortion. She is obese, and has always had irregular periods and so did not notice that she was pregnant until recently. The pregnancy was an unplanned consequence after recreational sex during which she was pretty sure he had used a condom. So, she visits her GP, the form is signed, a date is set and there she is at the hospital for the procedure. Because she is at 20 weeks there is no problem with the law. Also because she is at 20 weeks, the foetus has to be killed first and then 'delivered' by induced labour. The killing normally involves an intracardiac injection of poison (such as potassium chloride), followed by a time period to make sure that the heart has stopped beating. Aiming the injection needle through the woman's abdomen and into the foetus's heart is a delicate process for the obstetrician, and impossible without a powerful ultrasound scanner.

Now here's the question. The ultrasound is powerful enough to pick up many features of the foetus that make it look like a baby – not only the limbs and fingers, but above all the face: eyes, nose, mouth. That is precisely what happily pregnant women delight in when they come in for their normal 20-week scan. During the injection procedure it is only the obstetrician who sees the ultrasound image. But what if the monitor were turned towards the woman so that she could see exactly what was going on? And what if the obstetrician accompanied this with a live commentary of what he was proposing to do?

Let me pause right away, for this scenario is bound to outrage certain readers. Such a suggestion will seem the height of insensitivity, coming close to severe bullying. After all, the woman in question is probably going through an experience that is traumatic and humiliating, not to mention physically very uncomfortable; she probably feels vulnerable and lonely enough as a patient in a large anonymous hospital, especially if she is a confused teenager; the last thing she needs or deserves is to be implicitly accused of requesting the murder of an innocent child. Quite understandably, she just wants to get shot of the whole wretched business. Nobody who ends up having an abortion, especially for the first time, can claim indifference to the procedure; surely the obstetrician does not have to rub it in? (And of course I do not need to mention how much more insensitive such a move would be for a victim of rape.)

In most areas of healthcare, extra information about a treatment usually leads to more informed consent. But there are certain areas where patients have an obvious 'right not to know'. The results of a post mortem can confirm a diagnosis for relatives without a detailed description of how the post mortem was carried out (I discuss this in Chapter 9). The experience is traumatic enough for the family as it is, and they just want to get Granddad buried as decently and tidily as possible. It might seem that an abortion patient has a similar right not to know, since her main reason for coming to the hospital is to have a problem solved. She understands enough about what pregnancy and childbirth would mean in her life, and *that* is what she wants to eliminate, not the foetus as such, which she has never seen.

However, the whole point about the vegetarianism example is that some vegetarians feel very strongly that carnivores *ought* to see, and ought to try to understand what exactly they are doing, even if they choose to continue eating meat; to avoid that is to be culpably, irresponsibly blind in a way no different from a hit-and-run driver. Can the same be said for the woman requesting an abortion, or is it just another routine medical treatment that is pretty gory and she has a legitimate right not to know further details? I don't have an answer to that question here, and I will explain why I hesitate to even look for one in the next section. But I want to suggest that this is a much more philosophically interesting question than the general permissibility of the procedure itself. By saying this I am distinguishing two kinds of normative pressures. The first is the pressure to *do* the right thing or the best thing or even the least bad thing. The second kind of pressure is to have the correct attitude, where the correct attitude is compatible with subsequently taking any number of actions. This second kind is the normativity at stake in my mini-scenario. Here it makes sense for one person to criticise the other because they are not taking a serious issue seriously enough. They have a cavalier disregard for human life. The failure might be due to genuine blindness and a lack of thought, or it might also be due to a wilful self-deception.

But am I not begging my own conclusions here, in the same way that I accused mainstream philosophers of doing when they merely stipulated that the foetus was below the threshold of respect? For I am taking the early-stage foetus to have the full moral status of the adult into which it will probably develop if allowed to (based on my argument from potential), and so of course any failure to recognise that will be culpable. In addition, my whole approach has been to avoid abstract theory in order to focus on what people actually *do*, what people have 'made' of animals etc. Surely there are many people today who simply do not see early-stage abortion as an ethical problem; just as there are many people who have no ethical qualms about going for a fourth or fifth abortion; just as there has been no serious political challenge to the Abortion Act in the UK. Whatever the sociological explanations for this, whatever the judgement of individual conservatives that this trend might be a bad thing, it is hard to deny that the trend exists.

However, I do not believe that the wider UK public is as unanimous as the absence of organised opposition to the legal state of affairs suggests. I suggest that people still have doubts about abortion, and that these doubts are very much ethical in nature. This suggestion is an empirical claim, so I can only point to a possibly fruitful line of further research for social scientists – if a reliable and clear way of testing the suggestion could be devised. I believe that such research would show *not* that people believe that the foetus is just a bunch of cells, nor that it is a stranger whom they are not obliged to rescue, but rather that they do not really think about the matter at all.

The *Guardian* G2 supplement of 27 October 2006 featured a garish pink cover, with the words 'Abortion is not evil'. Inside, Zoe Williams writes: 'With pro-lifers dominating the debate, and even left-wingers describing abortion as a "necessary evil", women's hard-won rights could soon be under threat'. The article is revealing. It cannot be true that pro-lifers are dominating the debate if there *is* no serious debate – Williams does not quote any MPs or any senior medical consultants, for example. Instead, it turns out that the 'debate' lingers in the public attitudes to abortion, according to which, she says, it is 'an unarguable tenet of modern society that you should feel ashamed of having a termination'. Williams then asks why there are no *jokes* about abortion, if, after all, abortion is just like 'having a tumour removed', as she says. The rest of the article interviews some women who had experienced poor treatment while having abortions, allegedly because of the negative ethical attitudes of the clinicians involved. Presumably Williams would see herself as a campaigner in the emancipationist mould; but she would not be able to explain how on the one hand slavery has been utterly discredited, while on the other hand women continue to support abortion intellectually and yet feel shame when they have to have one themselves. Williams interprets this shame as associated with more general hypocritical attitudes towards female sexual behaviour and social roles, attitudes involving the tacit belief that only a promiscuous or heartless careerist would need an abortion anyway. But I would suggest that another component to the shame is the revelation of the ethical reality of a life callously destroyed, and that the foetus has not allowed itself to be so conveniently reduced to the status of a tumour or appendix. This is not to deny that woman should have the right to an abortion under certain circumstances; nor is it to deny that a woman can make a responsible and sensitive decision that it will be necessary to kill the child inside her. But is to deny, very forcefully indeed, that anyone should make jokes about abortion.

Perhaps turning the ultrasound monitor towards the woman is too drastic. Besides, at earlier stages of gestation there is not much to see, and even with good imaging technology what can be seen probably looks sufficiently inhuman to encourage a woman's belief that the being is some sort of parasite, fit to be killed. But the 'bringing to see' need not be so literal. Would there not be an argument that women requesting an abortion should at least have a *real* conversation about it first? What I mean by 'real' is something more than the carefully scripted performances by the woman and the GP, each going through the motions and saying what they feel they ought to say. And there is the same risk here as there is with the allegedly non-directive genetic counselling session; it is disingenuous to suggest that the GP can merely spell out the options. The problem is that most GPs have neither the time nor the training for such a conversation. It is a remarkable testament of the authority vested in 'registered medical practitioners' that *any* of them will do for legal purposes, regardless of the fact that the vast majority of them have had no training in counselling, psychotherapy, social work or working with distressed teenagers. Room should be made for such training in the medical curriculum. But even so, there is a real argument for amending the Abortion Act to bring its implementation under the power of social services.

At the very least, doctors can themselves engage more with the 'real conversation' by asking themselves how they live with the *contradictions* of abortion. The contradictions are like those in the stories about animals that we tell children. How do we explain to children that we stroke and cuddle these animals here, that we shoot and poison those animals there and that we eat those animals over there, when there is no obvious relevant difference between them. In a similar vein, how does a medical practitioner reconcile with himself his sincere joy at the news of his wife's pregnancy on the one hand, with the signature on the form that condemns another child to a premature demise on the other hand, *when both the beings in question are the same.*

The proposed conversation with the woman might still be too much for people like Thomson or Warren, who see the matter as of no business for anyone else but the woman. Designing a system that would require such conversations surely takes us back to the bad old days of paternalistic medicine, where patients were not treated as responsible adults. So yes, I am arguing for a paternalistic measure, not only on the basis that the doctor (or social worker) might know better what is in the woman's interests than the woman herself does, but also because the doctor has to act as an advocate of the 'second patient' – the child. Again, we are back at the beginning with the problem of seeing. Is it a child or is it a 'bunch of cells'?

There are some in the pro-life lobby who would argue: 'Fine, if you don't want the child or feel that you cannot handle parenthood right now, then give it up for adoption. There are plenty of people willing to give it an excellent upbringing'. Such an argument would not please Thomson, of course, who see the nine months as grossly excessive 'punishment' for an honest mistake, all for the sake of a stranger. But there is another point to make in response to this. Giving a child up for adoption is in many ways more difficult than having an abortion, for the apparently paradoxical reason that the child will be alive, and therefore alive to return one day and ask me why I abandoned him. With an abortion, I'm shot of the whole mess. 'It wasn't my fault, I didn't enjoy going through with it, but there was no way I could keep it, and now I'm moving on.' With the child alive, out there somewhere, I can never really move on because there is always the fear of the knock on the door when the child gets older. Again, though, this understandable reluctance to be in a situation where I will have to surrender the child is not a sufficient reason to have the child killed.

Proximity and authority

Let us recall Warren's claim that the threshold of respect should be at birth itself because the foetus/baby is then no longer physiologically dependent on the mother and can be given up for adoption rather than killed. I rejected this claim because I found it bizarre that a being might undergo such a massive transformation just by physically moving down the birth canal. This position is analogous to the curious part of the English law, according to which a 'severely handicapped' foetus can be aborted at any stage of gestation, but the same severely handicapped child, once born, cannot be actively killed (although it may be allowed to die). Finally, let us also recall Thomson's claim that the foetus is a stranger to the pregnant woman, and therefore that she has no more obligation to rescue it than to rescue other strangers when costly. I rejected that claim by arguing that the baby was of the woman's flesh, whether she liked it or not, and therefore that she bore a greater obligation to rescue it than she would to a stranger.

There is perhaps one aspect of the above three positions that holds water, and which I want to consider in this final section. Might it make a difference that the foetus is not *visible* to other people or even to the pregnant woman herself, and that this fact makes for a difference in the meaning of 'foetus' and 'child'? At face value, the suggestion is ludicrous: if I close the door to the room where my infant child sleeps, that hardly makes a difference to my obligations to it. But this analogy is imperfect because I already know very well what my child looks like, whereas I have never seen the foetus before. And yet the possibility seems to go against two aspects of the position I have been developing; namely, that (i) *even though* I could not see the foetus now, it would develop into an adult human being whom the contemplated abortion would effectively be killing and (ii) although the woman does not see her child, she certainly feels it kicking, and this leads to a much more intimate, visceral relationship to it.

Consider a related paradox in one of the guidelines to the new Human Tissue Act.[1] According to the Act, a woman who has an abortion can request that the 'foetal tissue' be given a proper funeral. So we can imagine a case where the woman considers the foetus to be somehow less than a full person, perhaps no more ethically special than her appendix, and so available for destruction on the slimmest of grounds before the 24-week threshold; and after the abortion, it suddenly becomes a full person, worthy of a coffin and prayer and perhaps even a eulogy. This might have to do with the shame that Zoe Williams wrote about, a shame that might even have surprised the woman when she suddenly realised what she had just done. Alternatively, the woman might have known full well what she was doing in asking for the abortion, and only did so with the greatest reluctance and regret, and the funeral was now the least she could do.

This notion of the foetus being invisible and therefore 'distant' ties in with other debates in moral philosophy about our obligations to others. Members of my family have a greater legitimate claim than total strangers; members of my ethnic group or nation have a greater claim than foreigners; a beggar in my street, here in front of me, has a greater claim than a beggar far away in a foreign country. And yet the strangers, foreigners and foreign beggars may not be different in any other ethically relevant way from those close to me; most importantly, their needs might be identical, or indeed greater. I spend money on my family while my fellow citizens go hungry in the streets. The NHS, partly funded from my tax payments, treats British citizens with minor ailments while millions starve in Africa.

If philosophers are not to condemn this 'principle of proximity' outright, they have to find ingenious ways to allow it into their theories. For example, rule-utilitarians might give preference to more proximal claims on the basis that everyone will be better off if everyone looks after their own. Others will suggest that there is a duty to develop one's talents and pursue projects and relationships that give meaning to one's life, for otherwise we would all have to aim to become Mother Theresas. But such attempts have always struck me as contrived and *post hoc*. Instead, maybe we should just accept that ignoring strangers is something that we simply do, and therefore trying to justify it philosophically is as fruitless as trying to persuade the Carnivore using arguments. It then becomes important that we cannot see the foetus, cannot look into its eyes, cannot caress it; as such it does not exist in a public space in the way full human beings do. We all 'know' there is a foetus inside the bump, but we have no direct experience of it, and therefore – in line with my arguments elsewhere in this book – it does not exist except in the abstract, and cannot ground ethical claims of the same strength as human beings with whom we can interact. (The ultrasound image is not a *direct* experience, since it is mediated through cables and electronics. For all we know, in the experiential sense I have been developing, the image could be animated by a technician in the next room.) Whether this distance then outweighs the proximity of the foetus as *felt* by the mother will be the question: but there might be this much truth to Warren's and Thomson's assumptions.

This notion of proximity has another, perhaps more controversial aspect to it. Recall my argument from Chapter 2 that philosophers had less authority to speak on matters of medical ethics because they were less proximate to the medical world than healthcare professionals; that is, they lacked the relevant experience. In a similar way, women are more proximate to the experience of pregnancy and birth than men are; and therefore, men have less authority than women to speak on abortion, either as a matter of policy or as a personal decision. This is a general position, and it will of course admit of exceptions; so the minimal claim is first, that the burden of proof lies with the particular man to demonstrate, somehow, that he has equal authority to speak on the matter as a particular woman; and second, that men would be advised to approach the matter with a certain humility; that is one reason why I have been reluctant to take up a more substantive position in my discussion. This might seem in stark contrast to the confidence with which I have been developing my position for most of the chapter. But again, I have been writing about certain women lacking a sufficiently respectful *attitude* towards the foetus; I have not been writing about certain women simply making the wrong decision or certain doctors committing a moral wrong.

Men are generally less qualified to speak about abortion because they will never have to make a decision about their own pregnancy. As such, their opinions come too 'cheap' since they will never have to pay for the consequences of their opinions. There is also a sense in which they do not quite know – once again in the deeper, more experiential sense of knowing – what they are talking about, cannot make the full imaginative leap to the woman's point of view, and therefore cannot sufficiently stand behind their opinion. Of course, one implication of my claim is that the Catholic priest has even less authority to speak on the question of abortion since not only is he a man, but he is a man who has abjured intimate relationships with women, abjured the life of fatherhood and family, and indeed who has abjured the life of the ordinary community of families; and so he does not sufficiently understand the stresses that children can place on a couple, let alone on a single mother.

I have been arguing that one first needs to understand something of a normal pregnancy and childbirth and something of a child that is wanted and loved *in order* to understand how a pregnancy might not be wanted; in a similar way, I am now arguing that women understand better what abortion is because they understand better what pregnancy, birth and childhood is. The foetus which they are considering aborting is of their flesh, beneath their ribcage, kicking. This is a far more intimate relationship than the father is capable of, even if he can touch and listen at the woman's belly. Whether or not a woman decides to abort the foetus, it is she who will have to bear the consequences of that decision. Even if she sought and followed advice, she knows that she *chose* to follow it, and so bears the ultimate responsibility. Even if she genuinely believes that the foetus is a stranger, or is only a bunch of cells, or is beneath the threshold of respect, it is she who will have to live with the regret of aborting it – if she regrets it, which of course she may not.

Even a *childless* woman, even a young woman with no experience of sexual intimacy, has a better understanding because she has grown up as a girl and a woman, that is, as different from boys and men primarily in her ability to conceive and give birth to children. She is also *at risk* of accidental conception through recreation or rape; her parents will warn her that it will be she, and not the spotty boyfriend, who will have the problem, and that is why she has to be careful in a way that teenage boys never do. The popular culture, and especially girls' and women's magazines, is full of images and ideas – implicit or explicit – of women's potential to become pregnant and to become mothers, willingly or unwillingly.[2]

Note that I speak of abilities, possibilities, images; I am staying well clear of any putative 'duty' to become a mother. Of course women should be able to follow career opportunities and choose childlessness if they please. My point is more about the intuitive understanding that every woman has about what her reproductive organs are capable of doing and what the implications are for her own life.

6
The Shape of a Life

In the last 20 years or so, there has been a great interest in the importance of narrative in understanding human lives and human actions. Belatedly, narrative has assumed its rightful importance in medical ethics as well (see Charon and Montello 2002, and Nelson 1997), but it has still not sufficiently informed the mainstream. Hence there is still the need in books like this to explore the main points, and again this will involve a substantial diversion away from discussions of medical ethics. Later, the pieces will be in place for a better understanding of old age and of the distinctive view of the lives that the elderly have. As such this chapter will serve as a bridge between Parts II and III.

The basic idea is that humans lead lives that are essentially extended in time (that is, diachronic), essentially organised into narratives, and essentially embedded in ongoing relationships with other humans and commitments to organisations or ideals. To repeat the metaphor that I have used elsewhere in this book, every human being is standing at a certain location, on his way from somewhere to somewhere else. This is in sharp contrast to the dominant conception of moral philosophy and medical ethics as essentially involving a synchronic problem, here and now, faced by an anonymous, solitary and abstractly rational agent. Such a conception is *atomist* in two ways: vertically, in its assumption that the individual has no essential relation to his past and future, and horizontally, in its assumption that the individual has no essential relation to certain projects and to other people. Vertical atomism leads to the familiar metaphysical dilemmas about personhood: is this person in 1980 the same as that person in 2007, even though they share the same (albeit changing) body? The standard answer, which finds its most detailed expression in the work of Derek Parfit, suggests that it will be the same person only if there is sufficient 'psychological continuity' between the two persons; roughly, that the 2007 person can remember enough of what it was like to be the 1980 person, and that there will be enough similarities in personality, interests, values etc. But such a framework of the metaphysical problem ignores the degree to which we are *treated* as the same person by others, especially significant others, regardless of our particular memories. This is one starting point for our discussion of what follows. In parallel, horizontal atomism neglects the degree to which we are *concerned* about what others, especially significant others, think of us and say to us: the hermit or existential hero or solitary Cartesian enquirer may attract philosophers interested in studying the pure human will or reason, but they are otherwise rare and exceptional figures in our ordinary world. It is this concern that I shall take as another starting point, for I will argue that it leads to a much greater integration between two people than the physical separateness of their two bodies would suggest. And this 'involvement' is almost metaphysical, having to do with attempts to define the self.

Both of these anti-atomist starting points invoke the notion of narrative structure; that is, the narrative structure of a single human life and the narrative structure of the relationships (with other people or organisations or ideals) that partly constitute it, for we live our lives *in medias res* (in the midst of things). In that way we may speak of a marriage, say, as having a story that is distinct from the two stories of its members, although there will clearly be a great deal of overlap. In addition, the narrative of a life can be revealed, to varying degrees, and to varying degrees of accuracy, to the onlookers: to the person whose life it is, to his significant others who see themselves as part of his narrative, and to third parties such as biographers, trying to make sense of the life and the relationships that partly constitute it. A narrative may be explicitly recounted, but it need not; it could just refer to the structure without a narrator. Most films and plays lack narrators: they seem to start as a chronicle, that is, as a loose series of distinct events, but the spectators are gradually led to assemble the narrative as the relationships and the relevant past details are revealed. One can also eavesdrop on a conversation in a train, without a writer's manipulative efforts.

Consider E. M. Forster's famous example: 'The queen died. The king died.' One corresponding narrative might then run: 'The queen died, and then the king died of grief', and already the notion of grief has temporal resonances beyond the present tense. Many different narratives, equally coherent, can be woven around a single chronicle (or, to reverse the order and use Hayden White's term, a single chronicle can be *emplotted* into different coherent narratives); one need only think of the many proposed narratives invoked by an imaginative detective when investigating the king's death. This does not mean that a chronicle is *wide* open to emplotment, however; for it has to 'ring true', given what ordinary people know about the world and about human behaviour, and given what certain individuals know about *this* person (e.g. 'Luther would never do that'); and one narrative, although possible, may well be less plausible than another. In the case of a murder investigation, there are also clear facts to which the narrative is answerable. If the queen turns out to be alive and well, then the king's grief was simply mistaken.

In hearing about the king and queen, not only do I come to know of a *causal* link implicit in the word 'grief', but it is more than that: the grief only makes sense as a result of the death, and understanding the nature of the grief means understanding something of the depth of the relationship that existed between the king and queen before her death. (Contrariwise, understanding something of the relationship will allow one to understand the grief.) The concept of grief, ascribed to the king in the present, thus necessarily refers to the past and future. Indeed, it might be impossible to fully capture the relevant aspects of that relationship without reference to the fact that the king's grief for his dead queen was sufficient to kill him; the grief *reveals* the essence of the relationship in a way that other descriptions – while she was alive – could not. Importantly, the grief may reveal the essence of the relationship *to the king himself*; indeed, he might be surprised just how much he grieves. In this sense he is not the full author of his own narrative. Importantly, narrative explanations can work in both temporal directions: another narrative of the above chronicle might see the queen as killing herself for the sake of her husband, either in self-sacrifice or to punish him.

This then links up with two notions I have considered earlier. Firstly, with the notion of bringing-to-see that I explored in Chapter 2. For in bringing the Carnivore to the slaughterhouse, the Vegetarian is trying to bring him to see the narrative connection between the animals' lives, their suffering, their deaths, and the meat on his table. More generally, trying to understand another person is often a question of seeing where they're coming from, and where they're going. Secondly, with the problem of different descriptions (with different resonances) that I explored in Chapter 3.

Consider a famous discussion of narrative in Alasdair MacIntyre's *After Virtue* (1985). There (p. 206) he offers us an apparently banal example of a man digging in his garden, and asks us what could be going on;

to make it intelligible we need to fill in the background detail, which will include his intentions of various scope and at various hierarchical levels. At the basest level of narrative, the man is *digging*, rather than putting a spade repeatedly in the earth; and the narrative end point will be a hole of sufficient depth. But beyond that his intention might not be to plant but to prepare the garden for winter, which invokes the narrative of growth and seasons; or to please his wife *by* preparing the garden for the winter. As an observer, I may only come to conceive the scene under one description, and it may not be the one running through the man's head or the one which you (keeping me company on this police stake-out) relate to me. But beyond that again, his intentions would refer to his ongoing gardening hobby or to the narrative history of this particular marriage. His marriage will itself be partly guided by interpersonal expectations of appropriate marital behaviour (i.e. the institution of marriage will have its own narrative history in that community). As MacIntyre continues,

> we cannot, that is to say, characterise behaviour independently of intentions, and we cannot characterise intentions independently of the settings which make those intentions intelligible both to the agents themselves and to others.

> (Ibid.)

Indeed, maybe the man will have two intentions, operating on differ- ent levels, or two overlapping local intentions with room for some give-and-take under a third overarching narrative, which, at the limit, will be the present phase of the story of his life. An observer would want to ask the man himself which intention was primary, without necessarily assuming that the man would be a full authority on his own intentions – after all, the man will also be a character in the story *he tells himself* about his past and future, and such a story will require some rationalisation to explain the mysterious or unsavoury parts and to preserve some measure of overall coherence in the light of the man's determinate ideals of behaviour.

This will not only be a description of his present commitments and memberships and where they're leading him, but most importantly a story of where he's coming from. The gardener must understand his past choices in terms of placing himself at different times in a certain orientation towards, and at an intelligible distance from, various more- or-less specifiable values and disvalues. He will incline towards one pos- sibility over another because he considers it valuable in some way. And each such placement must join up with the next to form a narrative route, and must join up with an expanding sense of the past. Part of my understanding of my own past will be regret over decisions I wish I had not taken; either because of some contemporaneous information that I now wish I'd (or now feel I should have) known then, or because of my present knowledge of the then-unpredictable consequences that that decision generated. Another part of my understanding will concern decisions which I felt then, and feel now, I *had* to do, decisions which in an important sense I had no option but to make, given who I remember being and what I remember mattered to me. Struggling to understand requires not only a sense of epistemic integrity – what makes him both a single interpreting subject, makes his interpretations and self-interpretations of a piece, and makes him the same person then as the person he is now – but also of narrative integrity, so that he can see his reasons *as* his reasons for tracing just that route and no other.

Dialogue

In my efforts to understand my life, I often choose to *tell* the story to another person. But such telling may not be a straightforward report of a finished package. Rather my understanding and my telling might be internally interrelated, for the telling might help me to understand the story better, especially if I am concerned about and surprised by the reactions of the listener. This is the phenomenon that Charles Taylor calls 'dialogicality' – human lives and relationships are essentially revealed and developed in dialogue between human beings. Hence my continuing demand, throughout this book, to know what people would *say*, to me as a philosopher, or to each other. Let me introduce the idea by discussing an example from Peter Goldie, where he is recounting the following mini-story to some friends in the pub:

> Last Saturday I went to the ground to watch the match, and stupidly bought a forged ticket from a ticket tout who seemed honest enough at the time. I ended up missing the game and trudging home fifty quid worse off, wet, angry, and thoroughly sorry for myself.
>
> (Goldie 2001 p. 13)

At the very least, this is a narrative rather than a sequence of thoughts in that the later events make sense in terms of the earlier events (assuming the unproblematic recognition of *those* experiences as *mine*). However, it is richer than that because in telling the story, I adopt a more objective retrospective viewpoint of greater ironic knowledge than that of the viewpoint of the hero of the story: for I did not buy a *'forged ticket'*, only a ticket that looked genuine but turned out to be forged. He did not *seem* honest to me at the time, I *took* him to be honest. The later knowledge of the tout's dishonesty 'contaminates' the narrative, says Goldie, preventing full imaginative identification with my former self. Note the different emotional responses associated with each viewpoint: *then* I felt angry and sorry for myself; *now* I see that my anger and subsequent self-pity were too self-indulgent, and that there are more important things in life to get angry about. The contamination works the other way too, since more-or-less articulate or conscious memories influence the present experience, as when one is 'once bitten, twice shy'. As such, remembering is as much of a reconstruction as it is an unearthing.

In telling the story I am watching the reactions of my audience to ensure that they are understanding what happened in the past; that they are making sense of my response *then* and of my response *now* and the ironic contrast between them; and that they come to *share* sufficiently my present viewpoint on what happened and to understand sufficiently both sets of emotional responses (i.e. that my anger seemed *appropriate* then, but in fact – i.e. in retrospect – was not, and that my gullibility is now mildly amusing). However, I may not succeed in getting them to understand or in drawing these responses from them; I may have to think quickly to come up with additional detail from the narrated events, I may have to explain some key links between the viewpoints, and I may also have to explain why it is supposed to be amusing (which, notoriously, will rarely ensure that it *becomes* amusing). Sometimes I might deliberately embellish, and recall the modification *as* an embellishment. Sometimes the distinction between the embellishment and the embellished may be blurred at either the moment of telling the story or in the version that I later remember.

This is very important. My attempts to make sense of a certain episode in my own life often rely on telling stories about that episode to other people, especially those whom I want to help to understand me and whose opinion of me I value. In this way I can 'try on for size' a tentative interpretation of the events and seek their reaction. Insofar as their opinion matters, I take the trouble to consult them, to talk things through with them, to defend myself to them, to defy them, to apologise to them etc. Rarely is this advice or criticism a matter of information; most often it is a special kind of emphasis, of phrasing, of placing into context, of trying to get the other to see. Indeed, 'trying

on' can be taken to extremes: as a teenager I can 'try on' whole personalities copied from films and refine them on the basis of the
response from significant others (and again, this need not be a fully
conscious, fully chosen process); adults can do the same, of course, e.g.
when it comes to exaggerating one's indignation because one feels it
is called for by the situation, by some person in the situation or by the
personality ideal to which one is striving. There will be pockets of
one's past that remain 'narratively opaque', in the sense that I cannot
weave them into any plausible story; and too many such pockets can
lead to alienation – I remember the events and my responses well
enough, but I cannot emplot them into a sufficiently plausible *and* sufficiently explanatory narrative, given my present understanding of
who I am now, who I was then, who I want to become now and who
I wanted to become then, and I say, helplessly, 'I don't know what got
into me'. Sometimes the past event may be so traumatic as to demand
opacity, as when the survivors of concentration camps fail to rebuild
their lives afterwards.

So narratives and stories have the particular sense they do only within
the context of being told to an *audience*, where an audience can range
from a packed theatre to a single interlocutor in an ordinary conversation, to oneself. The storyteller has to anticipate the audience's expectations and responses to some degree, and to fine-tune the telling
accordingly. Each audience will merit a slightly different version of the
same story, if only because the second telling will involve reflection on
the success of the first telling. Note that the word 'audience' should not
be thought of as a final, passive stage of the process of creation and
delivery: rather, the audience is actively engaged as one side in a dialogue, and such engagement requires a sufficiently shared perspective
and understanding. Even in the contrived environment of the theatre,
where the audience usually has no lines in the play and sits in darkness,
they still retain the capacity to surprise the actor, and there will always
be room for improvisation by 'feeding' off the audience's reactions.[1]
And as we have seen, the very process of telling the story will very often
influence the remembered content and meaning of the story. What
should be guarded against is the thought that the story is *there* to be
told, that the story will be unaffected by the telling, as a mountain is
unaffected by being painted well or badly.

When young, we learn a rich capacity for expression through language,
gesture, art and love, mainly in dialogue with our parents and relatives,
but then with teachers and priests and with fellow first-language-learners
too. So much of our self-conception – both in cool reflection and in

moments of crisis – is therefore wrapped up in formative dialogical exchanges from the past. 'Moreover', continues Taylor,

> this is not just a fact about genesis, which can be ignored later on. It's not just that we learn the languages in dialogue and then can go on to use them for our purposes on our own. This describes our situation to some extent in our culture. We are expected to develop our own opinions, outlook, stances to things, to a considerable degree through solitary reflection. But this is not how things work with important issues, such as the definition of our identity. We define this always in dialogue with, sometimes in struggle against, the identities our significant others want to recognise in us. And even when we outgrow some of the latter – our parents, for instance – and they disappear from our lives, the conversation with them continues within us as long as we live.
>
> (Taylor 1991 p. 33, italics in the original)[2]

Dialogicality therefore helps to anchor my self-understanding close enough to reality and to partly counter the fear that I might emplot my life into any number of wild self-delusions. My perspective on a situation is rarely the end of the matter; I have to deal with others' perspectives, and with what I imagine to be others' perspectives, all the time, since I am engaged in ongoing and essentially reciprocal relations and projects with them. I do not simply encounter the situation, respond to it with an evaluative description, judgement or action, and walk away. If the others are significant others, I have to take their judgements seriously; not only does the situation demand a response; *they* demand one. And if I care to remain as morally competent in their eyes (where I deliberately use the word 'care' rather than 'decide'), I must at least address their concerns and provide explanations of my actions which will satisfy them. Everybody but the radical sociopath will have intimate relationships, and it is partly within the context of their ongoing relationships (and in the context of their perspectival awareness of their ongoing relationships) that individuals develop in *this* direction rather than *that* one. To put it another way, I am never alone. The web of relationships I am involved in accompanies me wherever I go, and partly informs my perspective and my experience within that perspective.

Consider the power of a symbol like a wedding ring. When I wear such a ring, I can never be alone in a room again. Perhaps to some this will sound like relentless persecution, and to others like crass sentimentalism; for better or for worse, it is intelligible to say that my identity has become so intertwined with that of my spouse that I am not sure any more where I end and she begins. My enjoyment of the good things and my suffering in the face of the bad things in life will be given a significantly different texture insofar as they are *shared*. The dialogicality of relationships makes it easier to understand why betrayal and abandonment by friends are peculiarly awful experiences, beyond the disutility or abrogation of duty that might primarily bother the mainstream philosopher with his atomist assumptions.

Momentous decisions

Now that we have a better grasp of the narrative structure of lives and relationships, and of what it means to tell a story, we can start to consider what it means to make sense of one's life as a whole: for example, what it means to look back on one's life as one is dying. For death is not just a biological event marking the end of a biological life; it is also the end of a human life, and the end of the story. In looking back on one's life, there will be certain events that had a major impact on causally shaping the person, and I will be calling these 'momentous', as distinct from 'trivial' (to borrow William James's [1952] terms). Typical, though by no means necessary, examples would be marriage, embarking on a new career, and moving to a new country, where I shall be assuming that the momentousness depends on a whole-hearted and open-ended commitment, and that each was sufficiently successful for a few years – long enough for the initial decision not to be revealed as hopelessly ill-informed.

Momentous decisions are to be contrasted with trivial decisions, such as buying a new car. The two kinds of decisions differ in the different kinds of responses that make sense when examining the decisions in retrospect. When buying a car, it makes sense to ask whether the decision was sufficiently informed, rational, and above all, *correct*. Is it what I expected? Did it turn out to be better than what I imagine the alternatives to be? It also makes sense to regret such a decision after making it. With momentous decision, the situation is more complicated, as I hope to show. But the added complexity is not because momentous decisions are just complicated trivial decisions. It is true that we might *say* something like 'it was a mistake to marry Karenin' or 'I should never have left Milton Keynes', in the same way that we may say 'it was mistake to buy that Lada'. Nevertheless I want to argue that it does *not* make sense to evaluate a momentous decision in terms of rationality and correctness, either for the author of the decision or for an observer; similarly, it does not make sense to regret the decision.

Discussions of normative practical rationality typically involve an agent standing before at least two options, such that the agent ought to choose that option that he reasonably expects to maximise the satisfaction (and likelihood of satisfaction) of his known desires/preferences (with trade-offs between the short and long term). Let us say I start thinking about buying a car in 1990; I then decide to buy one in 1991. I calculate how much I can afford, what sort of car I can get for that price, what I need the car for etc. In 1992, I have been driving the car for a year, and am thus in a position to evaluate my 1991 decision. In 1992, I can also imagine the possible world in which I had not bought the car, had the extra money to spend, but had to rely on the train to get to work etc. The 1991 decision can be correct in two ways. It can be *substantively* correct if the car turns out to be exactly what I expected and no further facts have come to light that would undermine the information on which I made the decision; it can be *procedurally* correct if it was made on the basis of the correct procedure (e.g. of gathering the right sort of information from sources of appropriate authority), but nevertheless the car did not turn out to be what I expected. I am not blameworthy for a procedurally correct but substantively incorrect decision. This framework is clearly simplistic, but I want to accept it as sufficiently precise for my purposes. For trivial decisions, such as that between the car and the no-car option, I shall not be challenging this framework.

Now momentous decisions differ from trivial decisions in that they have a big impact on the author's life (or rather, any impact that a trivial decision has will be accidental, and the author cannot be held responsible for it). A more important definitional requirement, however, is the *attitude* with which the sincere momentous decision is made: it must involve a whole-hearted and open-ended commitment to a person or project. We could summarise this attitude as follows: when I make a sincere momentous decision, I give it my best shot. Of course, 'best' here does not have any necessary link to ethics or to ethically admirable behaviour: Michael Corleone in *The Godfather* gives his best shot at taking over the New York mafia. The commitment could be explicitly lifelong (as in my marriage to Karenin), or it could at least allow for the real possibility that I may end up spending my whole working life in medicine or in Madrid. Importantly, there is a sense that a momentous decision is not entirely mine, and that I am at least partly *chosen* by the partner, the job or the country. This is reflected in the difficulty of providing satisfactory reasons for making the decision.

Let's now look at my momentous decision to marry Karenin. We meet in 1990; we marry in Bognor Regis in 1995; we give it our best shot, but we divorce in 2000. Was the 1995 decision correct? In a very important sense, it depends *when* you ask me, at what stage of my life. If you ask me in 1995, I'll bless the day. If you ask me in 2000, I'll curse the day. If you ask me in 2005, my attitude has become bittersweet: 'we both made mistakes', 'it wasn't meant to be' etc. The correctness of the decision to marry will depend on the significance of that day, and that significance may well shift, again and again, through the rest of my life. It is thus hard to speak of the decision being correct or incorrect, 'all things considered', or 'once and for all', as it is with the car. This is because the significance of a momentous event will depend on the context within which the later person evaluates it, and this context will include the accumulated experience between the event and the moment of evaluation. As with my caveat about emplotment, above, this lack of a fixed significance does not mean that it can signify *anything at all*, or that I can *make* it signify anything I like. First, it has to be answerable to the facts: it is either true or false that he was wearing a white suit, regardless of my memory of the tweed – and look, here's the photo to prove it. Second, it has to cohere relatively smoothly with general patterns of human behaviour. Third, it has to cohere relatively smoothly with my other memories. Fourth, it has to cohere relatively smoothly with the narratively structured memories of Karenin and/or of other observers with whom I am still in contact. But all these are limits that constrain, without determining the significance, and without determining the particular future shifts in significance.

When I said that I cannot make it signify anything I like, that is not quite true. Mostly I discover the significance and the shift in significance; I may be surprised to discover that the statement 'it's Karenin' or 'it's my husband' no longer carries quite the same weight in my deliberations. But I can try to *nudge* the significance one way or another depending on a higher order attitude to the significant object – I can choose to give or withhold the benefit of the doubt, for example. Similarly, if I treat Karenin badly, you might remonstrate with me by saying 'for God's sake, that's your husband you're talking to'. In terms of the propositional content alone, such a statement would be very bizarre; for I know full well that it is my husband. But the remark should instead be interpreted as an effort to remind me of something I once felt for my husband and of the ideal of marriage I once believed in. As such it is another attempt at bringing-to-see, and can be followed by a conscious decision to see him in a better light, that is, to nudge the significance.

This shifting significance will worry many philosophers. If there is no fixed and no correct significance, then surely it ultimately doesn't matter what the significance is; for there is an essential arbitrariness about it all, and the agent might as well not bother seeking significance. This point is well taken *if* one accepts the supremacy of the perspective from which it is made, that is, from an impersonal or third-personal perspective: the significance that events end up having can often seem quite mysterious and unpredictable. However, from within the agent's perspective, the events *do* matter, and the person cannot help but take their discovered significance seriously. When I look back upon my life, the wedding with Karenin stands out as a glaring event precisely because it led to five years of married life; its impact on me was such that I can no longer clearly imagine what my life would have been like if I had married someone else. Even if the significance shifts, even if it is not clear what the significance is at a given moment in the future (i.e. 'I don't know how I feel about him now'), there is no room for doubt about whether the event has a determinate significance. And given that human lives are essentially dialogical, the significance of an event, while perhaps obscure to me at the present, can be *jointly* revealed through discussion with a trusted friend, especially one who knows Karenin well. I can ask the friend, for example, whether I am being fair to Karenin, and whether he merits the benefit of the doubt: again, this reveals a genuine concern for an objective significance, a concern that allows no room for suggestions of arbitrariness. This is compatible with the significance of the event shifting in the agent's understanding as well as in the friend's, for time has passed for both of them, and there is no more objective perspective from which to assess which shift of significance is correct.

If I am to summarise the impotence of the philosopher's third-personal concerns about the arbitrariness of the significance, I would say it was because the philosopher has failed to understand what it means to *lead* a life from the inside. To illustrate this failure, consider the prenuptial contract, which has the force of law in some American states (although not in Britain). The idea is familiar enough, with a text running somewhat like the following: 'In the event of divorce, each of us will keep what we brought to the marriage, and will keep 50 per cent of what we acquired once in the marriage.' Prenuptials make a lot of sense, given the divorce statistics – about one in three marriages fails in Britain. I would be very foolish not to insure my house if there were a

one in three chance of it burning down. And however confident we are
that 'it can never happen to us', people change (as I have been stressing
in this very chapter) and grow apart, and unexpected adversities and
opportunities will bring out less appealing shades of character. Any
divorce is going to be unpleasant enough, say the lawyers, so why not
at least simplify the division of property. Besides, in the best case the
signed contract might never need to leave the bottom drawer.

Prenuptials make a lot of sense *from a third-person perspective*, that is
from the perspective outside the marriage. However, taking a prenuptial
seriously from a first-personal perspective means deliberately allowing
for the possibility of failure, before even entering the marriage. If my
marriage is to be serious, I have to have the appropriate attitude: one
crucial condition for the decision to be called momentous is for there to
be no room in the foreground of my (first-personal) consciousness for
the possibility of failure: that is what the decision to commit, the decision to trust, *means*. I cannot say 'as long as we both shall live' and 'let's
see how it goes' in one breath; I have to resolve to make it work. I may
be apprehensive, of course, but such an apprehensive awareness has to
be pushed to the background on the wedding day. On the other hand,
the *observer* can say 'let's see how it goes', for he is not implicated.

It would be tempting to accuse the 'uninsured' bride of some form of
hopeful self-deception or wilful ignorance. But this charge is not immediately clear, for the bride does not *deny* that one in three marriages fails.
The alleged self-deception or wilful ignorance in question is not really
about the truth of a proposition, but rather about the role such a proposition should play in the agent's deliberation and attitude. If someone
makes a decision to trust (and to declare that she trusts) someone
enough to be her husband, it would be a sign of pathological jealousy
and absurd control freakery if she then verified her husband's every
move by hiring private detectives to trail him. Yes, there is a possibility
that her husband might cheat on her – it happens. But who would want
to live in a marriage where that thought was in the foreground?

The change in the person

One of the assumptions of the practical reason model correctly invoked
to describe car-buying was that the person going into the deliberation
and the one coming out were the same – except for the one desire
satisfied, if indeed it was satisfied. Indeed, this is not usually considered
worthy of being an assumption, but is taken as just obvious. Any judgements about the correctness of a trivial decision presuppose a constancy

in the desires and preferences which the decision is meant to maximally satisfy. And even when my preferences do change over time, as when I grow bored of my Lada and start to yearn for a Trabant, this does not involve a change in *me*. The correctness of the purchase will then depend not only on whether the car turns out to be what I expected, but also on whether it turns out to satisfy my preferences: maybe it is only after driving the new Trabant for a while that I discover that my yearning was misplaced.

However, the practical reasoning model normally considers a brief period of time between decision and evaluation. With momentous decisions, I am talking instead about years, even decades. And I want to contrast one obvious claim – that people don't change in the short term – with another obvious claim: people do change in the long term. Momentous decisions, because they involve a profound commitment, will inevitably change the person, change his priorities, values, preferences and ideals. Not entirely, of course, and not right away, but recognisably to others and probably to the agent himself. This is the mirror image of my earlier claim that the significance of an event will change in time.

For example consider the statement: after ten years of working as a doctor, I have *become* a doctor. This is not a truism; nor does it refer to a set of knowledge and skills; instead, it is about the mindset, which has been influenced by the nature of the job, by the nature of my encounters with colleagues and patients the opportunities for advancement and achievement, and for pride and shame. I may resist some of this, of course, but even the resistance will be conditioned by my new environment and by the personal investment that I made into that environment when I first began. Again, I am assuming that my initial momentous decision to become a doctor was sufficiently well informed and lucky for me to give it my best shot, and to succeed in the profession for a few years at least: as such, even if I am bored or fed up with being a doctor, there must be something that attracted and maintained such a serious commitment to it.

However great the changes, it is worth stressing that the person remains the same in virtue of leading a single life, and of taking (or at least being expected to take) responsibility for his past decisions. In so doing I am rejecting the claim (often called 'existentialist') that one is not strictly bound by one's promises, but only *chooses* to keep them. This does not mean, of course, that I am bound indissolubly by my past promises; only that I will need a good reason to break them if I want to continue being taken seriously and sharing the same ethical universe as those whose opinions I value.

This is then a paradox at the heart of this chapter: how does a person stay the same (in virtue of leading a single life and in taking responsibility for past decisions) and change (in virtue of making momentous decisions)? One curious result of the changes, however, is that the person who *makes* the momentous decision will often not be exactly the same as the person who *assesses* it years later. Similarly, the people who observed the agent's momentous decision at the time will not be the same as those who assess it years later. Everybody has moved on. And so the criteria and standards that each would invoke when attempting to evaluate the decision have also changed.

This brings me back to one of the themes from Chapter 3, where I suggested that a woman might have to 'learn to love' the child of an unexpected and inconvenient pregnancy. To judge a decision to be 'incorrect' means that I have a clear idea of what my expectations were and that the chosen option did not meet my expectations or that the unchosen option has been revealed as more likely to have satisfied my expectations. With a car, this is straightforward and the unchosen option readily imaginable. But *ex hypothesi*, I only get around to evaluating the momentous decision several years on; by then it is so far in the past that I may no longer be sure what my expectations were, whether the expectations were reasonable or appropriate and in what way were they or were they not satisfied. Consider the distinctive perplexity involved in re-reading old diaries. There is no doubt that *I* wrote these words, and they are written in intelligible English – but what do they mean, and who the hell wrote them? Consider deciding to have or keep a child: there is no way that I could have expected, then, to be talking to *this* child. Can I look back on that decision and call it correct or incorrect?

The further away from the decision I have moved, the more difficult it is to imagine the counterfactual life-path. If I had stayed at home instead of moving to Madrid, any number of things might have happened to me and to my environment, some of which would have forced further momentous decisions. So after several years in Madrid some of the possible counterfactual life-paths would be unrecognisable to the point of preventing any real comparison with the actual life-path – and therefore removing any concrete sense from a judgement that the move to Madrid was a mistake. For all I know, staying in Milton Keynes could have been even worse.

What we are left with is a brute fact of the decision. I chose medicine rather than accounting, and as a result I am here now. I still have the choice to leave medicine, of course, and I may have the choice to retrain as an accountant, but this will be a very different decision than the one

I originally faced, since I am now a doctor. So the problem of evaluating the past momentous decision is not contingently difficult because of, say, a poor memory. Even if I had a perfect memory of the considerations that weighed with me when I chose medicine over accounting at the age of 18, I am now thinking as someone who has spent 20 years as a doctor. I can no longer get into the perspective of the 18-year-old I once was. And I cannot intelligibly regret having chosen medicine, in the sense that I cannot wish that I had chosen otherwise. This is not because what's done cannot be undone, or that regret is pointless. Rather, what I am regretting is actually that I am a doctor *now*, that it has ceased to delight me in the way it used to, and that my options have narrowed, and I am too old to start again. It is too late to regret the momentous decision to study medicine, because that is as much a part of me now as other brute facts, such as my family and ethnic memberships. The full concept of regret involves an educational component: I can learn from mistakes. But I have spent too much time being a doctor, and I have invested too much into it, for me to now consider it a mistake and to somehow undo that commitment. Indeed, can any 18-year-old know enough to make a rational decision about the rest of his life, as opposed to just stumbling into something for reasons that will never become clear?

Old age as the last chapter in the story

Part II began with pregnancy, birth and childhood. Most of this chapter has been interested in an indefinite point during a person's adult life. The rest of this chapter will concern the end of the person's life and the particular perspective on one's life that the elderly (or younger people with a terminal illness) are forced to adopt by the awareness of the end. For most discussions of narrative rightly emphasise the person's ongoing life, with a future that is essentially open. Of course, the future is not wide open, we know we're not going to live to be 200, but the end is far enough away that it is not *experienced* as imminent. Thus most of us know that we are mortal, but we do not really know – in the sense of the two kinds of knowledge we have been discussing throughout the book. Knowing that the end is nigh has nothing to do with the number of days or weeks; but regardless of where that knowledge finally comes from, it has a crucial impact on the way the elderly person will see his present and his past. Certainly, there are practical preparations and farewells; but more importantly there is also an attempt to make sense of one's life as a whole, that is, to write the 'last chapter in the story', to borrow Howard Brody's term (2003 p. 254). As any book-reader knows, the last chapter is not merely the final series of events, but it wraps up the story, and may do so more or less satisfactorily. No matter how enjoyable the story might have been to read through chapter by chapter, it is possible to *spoil* a book by a bad ending. It is the last chapter, and the effort to write it, that best characterises the elderly.

Despite their enormous diversity of personalities and backgrounds, the elderly are all too often united in healthcare policy discussions into a single class on the basis of their common biological ageing patterns, sometimes called the 'deficiency model'. Certain physiological functions decline in a given individual in predictable ways, and it will require resources to care, treat, and manage the decline as best as possible. Old age is a mirror image of infancy because of the return to physical vulnerability and dependency, and in some cases to the same kind of incompetence. However, the elderly are radically different from other patients with the same deficiencies because of the peculiar vulnerability to humiliation associated with a *decline* from a full and active adult life, and to humiliation associated with a spoiled ending to the story of their life.

Recall my discussion of shifting significance from earlier. The last chapter involves a final effort to fix the significance by considering the whole picture in a way that one was not forced to do earlier. It is worth stressing that such a last chapter could be positive or negative in tone: I could recall an event and forgive my enemy, or I could send him back to hell with renewed vigour. There is no necessary reason why the deep awareness of one's own death should inspire humility and charity over smugness and *Schadenfreude*. In the same way, I can compose the last chapter in a lucid and honest attempt to do others justice, or I can rationalise my own failings and guilt into a hollow greatness, or I can collapse in despair at how little my life has amounted to.

Ronald Dworkin (1993 p. 212) considers the subtle difference between a dying patient holding on *until* a particular event and holding on *for* that event. The former attitude is merely temporal, and involves enduring in one's struggle with a deteriorating body. The latter attitude confirms 'the critical importance of the values it identifies to the patient's sense of his own integrity, to the special character of his life'. If I have always been attached to my family and to family events, for example, there will be a special urgency in being kept alive long enough to learn of the birth of a healthy grandchild, or even better, to meet the grandchild and to bless him. Of course the person had experiences of family life that were enjoyed, appreciated and recollected in their own right.

But it is this last wish which ties together all the individual experiences into a satisfying ending of the story of a particular person, a person partly defined by the long-standing value that his family played in his life. The longing to hear the news of the healthy grandchild naturally emerges from the recollection and reappraisal of all the events in the person's relationships with his family.

Composing the last chapter is all about a search for meaning. It is not about remembering happy times or quality of life. In this respect the exclusive emphasis in medicine on quality of life (and its quantification) may be misguided in this context. In a similar way, the whole philosophical debate about whether quality of life is objective or subjective is missing the point. There is of course something to be said for a doctor to be interested in more than physiological performance indicators and mere longevity of life when comparing different treatment options. But my question is whether doctors also have a responsibility to think about the meaning of lives near the end, and of the particular damage that institutional indignity can do to that meaning.

For consider two banal facts. First, one can have a life of very good quality that consists of lazing about on the beach all day, living off an inheritance. And yet such a life can also be desperately trivial. Second, one can have a pretty miserable quality of life, struck down by illness and penury, and yet accomplish something of great worth: the lives of some great composers or writers or saints leap to mind.[3] When it comes to looking back over one's life, it will surely not be enough to reflect on how much quality or how much happiness there was in it. When I reflect on my friendship with Pantagruel, I can cherish the friendship, and I can cherish Pantagruel; but if it is a real friendship I cannot say that it was valuable only in allowing me to experience a happy and high-quality life. The friendship may have had its good times and its bad times, but whether on the whole it had sufficient quality – what on earth could that mean? There may have been occasions where I felt Pantagruel was exploiting my friendship, or was going cold on me, and I may have had to give him the benefit of the doubt, but in the end our friendship lasted, and looking back on it I now cherish it: what *more* can be said about the friendship than to describe it in those terms?

All too often, quality of life is spoken of as something that can be acquired, even purchased. Certainly the beach holiday, one supposed example of a high-quality life, can be purchased. But then it comes to resemble the pursuit of happiness. Imagine a case where I have purchased quality of life or pursued happiness at the cost to someone else. Now imagine that I am in the twilight of my life, looking back over it, and I come to truly appreciate what sort of cost has been incurred by this someone else. This fact undermines the putative quality or happiness that I only *thought* I had achieved. It is not as if I have achieved the quality, enjoyed it, and then, as in the song, 'they can't take that away from me'. The significance of the quality has shifted, and now strikes me as grossly self-absorbed, and the neglect of others now strikes me as profoundly unjust. Importantly, injustice in this context is not to be reduced to enlightened self-interest, as when I say to myself in my old age that I ought to have been nicer to my daughter so that she would look after me now. Rather, injustice involves the revelation of another human being as having a legitimate claim on me, a claim that I neglected, regardless of whether that neglect is costing me anything, now or later. The remorse I feel is not the result of poor calculations in the negotiation and renegotiation of the 'contract' at the time, but as a result of having wronged another. The experience of remorse is not something to do with happiness or unhappiness, or with a good or bad quality of life; it has to do with meaning and understanding. It can therefore be essential to writing a coherent and lucid last chapter.

Part III Matters of Life and Death

At the end of the last part, Chapter 6 acted as a bridge between Part II and Part III, between the broader topics of birth and death. It examined what it means for a human being to lead a life, and to have a unique perspective on that life. This examination serves as an essential background for any attempt to make full sense of the topics in this third part. In each chapter, I again criticise some mainstream accounts for failing to do justice to the ethical complexity of the situations and, in particular, for neglecting the afflicted individual's perspective on that situation and on himself in it. I also argue against the unquestioning application of certain key philosophical concepts, such as 'rationality' in the discussion of suicide in Chapter 7, 'personhood' in the discussion of dementia in Chapter 8, and 'autonomy' in the discussion of euthanasia in Chapter 10. In place of these concepts, I argue that certain other concepts, namely those of horror and pity, are more relevant in our efforts to make sense of what is going on.

Chapter 9 continues Chapter 2's examination of the extraordinary nature of the medical world by looking at the ambiguous status of the human body, both as a sick living body and as a dead body, and concludes with a discussion of the Alder Hey scandal. If the body is no more than a discarded shell, as many believe, why would it be controversial to secretly remove organs from it?

The final chapter examines the most difficult topic in medical ethics, and also the one that has rightly attracted the most attention from mainstream philosophers and in the popular press, that of euthanasia. Building on the discussion of suicide in Chapter 7, I consider some of the well-known arguments for and against, together with the famous English legal cases of Ms. B and Tony Bland. As elsewhere in the book, my aim has not been to defend one 'side' against the other, nor to arrive at a clear conclusion that could guide policy, but merely to appreciate the irreducible and mysterious complexity at the heart of such cases.

7
The Problem of Suicide

In the previous chapter I discussed the concern people have for the shape of their lives, that is, for the narrative coherence of the objective meaning of the events and relationships that make up their lives. It may happen, of course, that a lucid and honest review of one's life may lead to the conclusion that it must not continue. Traditionally, and among many people today, the desire to commit suicide can only be irrational, at best serving as evidence of a mental illness, at worst as culpable ingratitude for God's gift.[1] This has been countered by a modern sentiment that it might, in some cases involving terminal illness and/or great unrelievable suffering (or the prospect of great suffering), be rational. Either way it is worth examining because it reveals an important dimension of the meaning of death in our society. It will also be essential for our discussion of euthanasia in Chapter 10.

David Hume was the most famous modern philosopher to support the conceptual possibility of rational suicide, but the idea goes back at least to the honour-suicides of the ancient Romans (not to mention Socrates). More recently, Richard Brandt (1975) has been an articulate defender of Hume. In this chapter I want to start further upstream, as it were, from Brandt and Hume, by looking at the concept of rationality that they invoke as if it were self-evident. After examining cases where the concept is typically used, I shall argue that it cannot be applied to – or withheld from – the act of suicide or its perpetrator. The more appropriate concept in response to news of a suicide, or to a declared desire to commit suicide, is horror and pity, together with a humility in the face of the profound mystery at the heart of any such an act.

Brandt's argument is a utilitarian one, according to which the rational agent could make a sufficiently informed comparison between the likely utility of two possible futures – one future with him surviving, almost certainly in his present or worse state, and one without him – and make a choice that would be intelligibly rational. Utility was broadly construed to be able to allow for altruistic utility to others ('not being a burden') and self-interested utility for the agent himself; but the drive to promote either could be appreciated as rational. By 'appreciated', Brandt would mean that *both* the agent *and* any observers could in principle agree in making sufficiently reliable estimations and comparisons about such utility. And if the decision can be appreciated as rational, certain ethical positions would then follow: (i) the decision to commit suicide could and should be respected; (ii) it would be ethically permissible to provide assistance. Brandt is more interested in (i), although many in the debate surrounding euthanasia (or more precisely, physician-assisted suicide) have argued for (ii), on the basis of exactly this logic.

Is it so obvious what it means to be rational? Rather than seek a dictionary definition, let me consider some ordinary cases where the concept can be confidently applied or withheld. Recall once again the disagreement between you and me about the number of chairs in the next room, or about a capital city of a country. When we check an atlas, at least one of us will be proved wrong, and there will be nothing left for that person to think but that he was wrong: it would be irrational to refuse to do so – *unless*, of course, there was greater reason to believe that the atlas was mistaken or falsified (in which case an explanation is owed to this effect), or unless you were trying to wind me up. We might of course say, with Descartes, that outside deductive and mathematical reasoning there will always be some room for doubt, and the answer will always be a question of probabilities. But our society would never be able to function if people did not agree about the rationality of believing in a great number of propositions about the world on the basis of necessarily imperfect evidence or authority. Even something as narrowly specific as our criminal law relies on a widespread understanding of 'reasonableness' in its definitions of, for example, recklessness, culpable ignorance and the duty of care; and proof of criminal conviction must be 'beyond reasonable doubt'. In such a context, we can say that it is rationally, even if not strictly logically, *necessary* for you to believe that, say, Prague is the capital of the Czech Republic.

To repeat a point I made earlier, we can follow Stanley Cavell (1979) and say that it is only against the background of such widespread implicit agreement about correct, non-otiose ascriptions of the concept of rationality that there can be meaningful *dis*agreement about whether a specific course of action is rational or not. When I accuse you of being irrational to go bungee jumping (or smoke, or not wear a bicycle helmet), my accusation and my concern make sense to you, even if you disagree with the judgement. An essential part of being rational is to have sufficient knowledge of and concern for one's own interests, and to take relevant and sufficient action to protect and promote them (e.g. by avoiding risks of sufficient probability or severity). But there is enough leeway in all the above uses of the word 'sufficient' to allow for a constrained variety of attitudes to risk, all of which can still be intelligibly rational. Importantly, the same can be said about attitudes to one's comfort, appearance, and health.

The constraints are important: someone who wants to go bungee jumping without a bungee cord could be plausibly suspected of simply not understanding the risks involved or of being self-destructively irrational to the point of mental illness – or of committing suicide. This is similar to someone who insists that the atlas is incorrect without offering further explanation of why he thinks so; such a person, as Cavell puts it, has forfeited his right to be taken seriously in the community of reason. So it is important to contrast two kinds of accusation here: on the one hand, if you refuse to believe the atlas without further explanation, I am making a *deep* accusation of irrationality. If you insist on going bungee jumping (with a bungee cord this time), I will still make a *shallow* accusation, which reveals more about our differing attitudes to risk (comfort, appearance, etc.) than any truth about your greater or lesser degree of rationality. Clearly Brandt is interested in deep accusations.

To focus our intuitions, let me start by considering the example of Ken Harrison, the hero of Brian Clark's famous 1972 television play (and subsequently a film) *Whose Life Is It Anyway?* (Clark 1989). Harrison was a sculptor before an accident left him a permanent tetraplegic. The play is about his efforts to prove himself sufficiently competent (i.e. rational, informed and free) to legally refuse all treatment (including nutrition and hydration) and thereby to be allowed to die. Here is Harrison before the judge:

Judge: Alright. You tell me why it is a reasonable choice that you decide to die.
Harrison: It is a question of dignity. Look at me here. I can do nothing, not even the most primitive functions. I cannot even urinate, I have a permanent catheter attached to me. Every few days my bowels are washed out. Every few hours two nurses have to turn me over or I would rot away from bedsores. Only my brain functions

unimpaired but even that is futile because I can't act on any conclusions it comes to. [...]

Judge: But wouldn't you agree that many people with appalling
physical handicaps have overcome them and lived essentially creative, dignified lives?

Harrison: Yes, I would, but the dignity starts with their choice. If I
choose to live, it would be appalling if society killed me. If I choose
to die, it is equally appalling if society keeps me alive.

(Clark 1989 pp. 73–4)

These arguments seem strong, and in the play the judge is persuaded.
I shall assume that Brandt would also consider Harrison's a rational decision. The argument here relies less explicitly on Brandt's utilitarianism
than on a liberal respect for patient autonomy, combined with the idea
of self-ownership: the patient's body is his to dispose of as he pleases.
But in order to do what he pleases with his body, he has to demonstrate
that the possible future with him continuing to live is objectively the
worse option. The judge's point about the other people with handicaps
is important, however, for it means that Harrison's decision lies within
the limits of rational variability. A reader can still be sympathetic with
Harrison's plight, while not personally endorsing his decision. It is not
accidental that Harrison was a sculptor, and therefore partly identified
with the work of his hands, work that is no longer available to him: perhaps a reader who identified more with mental work might find
Harrison's situation more bearable (e.g. Stephen Hawking). On this
'liberal' conception of the disagreement, the decision would ultimately be
a personal one; any accusation of irrationality against Harrison would
be shallow, and his choice therefore ought to be respected.

Against this liberal understanding of Harrison's situation we have the
deep accusation of permanent or temporary irrationality, an irrationality not to be properly engaged with (i.e. one human being to another)
but only *treated* (the competent doctor treats the incompetent patient).
For rationality is all about protecting and promoting one's needs and
interests: and how could it then be rational to destroy the very bearer of
those needs and interests, and to destroy the possibility of protecting
and promoting them in the future? Any life is always better than no life
at all, so the argument would run. I shall not be taking this absolutist
position as a serious rival to Brandt's. It is similar to saying that no
smoker can *truly* understand what cigarettes do to human lungs, for
'otherwise they surely wouldn't do it'. (Again, such a claim is easily dis-
proved by the number of rational doctors who smoke.) However, there

are two important kernels of truth in the absolutist position, to which I shall return.

Horror and pity

It is hard to describe the great and distinctive unease associated with hearing news of another's suicide, an unease that I claim to be lacking in hearing news of an accidental or natural death; perhaps I can only appeal to the reader's shared experience. Some might think the dominant emotional response, when there is one, is deep sadness rather than unease. This might be true in certain cases, especially where the victim is well known to the respondent, i.e. where enough is known about the unhappy antecedent circumstances to 'focus' the respondent's sadness, as it were. But upon hearing the news of a *stranger's* suicide, I would suggest that alongside any response of sadness is also this distinctive kind of unease. Indeed, I want to suggest that it is a kind of fascinated horror. Consider the grim curiosity with which we enquire about the manner of suicide, as if it were the final revelation of the victim's character, the last chance to get to know him. This curiosity is similar to that which inclines some of us to watch 'slasher' films, or to pick at a particularly ragged scab. But there is a stronger reason to call this response one of horror.

By the end of the play, Ken Harrison is no longer a stranger to us. We have come to appreciate the circumstances driving him to suicide, and we certainly feel sadness at his plight. But there is room for horror as well: in the last scene he is not dead or dying as in a standard dramatic tragedy, *he is very much alive*; indeed, given the titular question of the play, it could be said that he has never been more alive, more aware of who he is and of what he stands for, never more articulate and passionate than in his arguments before the judge. He is striking to behold. And yet this clear-eyed rationality, this sheer defiant force of will, is what is ultimately so unsettling, so horrifying, because as we leave the theatre we know which way the story's headed. It is in the face of this horror that *our* discussions about the putative rationality or irrationality of Harrison's act *run out*; any ascribed rationality does not reassure us, for we know that his rationality will not get him anywhere, not achieve anything positive, will not expand his possibilities in the future. This is not just the awful bare fact of being alive, and then not; it is the fact of being alive, and then deliberately bringing one's whole world to an end.

To illustrate this last point about not getting somewhere or achieving anything, let us consider an ordinary, uncontroversial case of rationality, and an ordinary response to someone else's rationality. After studying the trends in housing prices, the neighbours sold their house just before the local market collapsed, and they moved to France. Perhaps I am happy for their good fortune or financial acumen; perhaps I am envious, and hope they discover rising damp and rats in their new French house. Either way, I implicitly acknowledge that what they did was rational (the planning and the decision) and that they have more money now than if they had waited longer. Now an essential part of the meaning of rationality, of having acted rationally, is that the neighbours will be better off *for the future*. Rationality – or at least this prudential aspect of rationality – is essentially future oriented in this way. I can admire their future activity or envy the range of their future choices. The rationality of their act, therefore, conceptually presupposes a future in which to enjoy the benefits of the act. Contrariwise, an irrational act derives its disvalue from the suffering that will most likely result in the future: I can admonish my other neighbours for sentimentally clinging onto their car despite the ever-increasing cost and frequency of repairs. My admonition is implicitly about their future and the opportunity costs incurred by their irrational choice today.

I said above that the 'absolutist' position – that suicide was essentially irrational – had two kernels of truth, and this is one of them. Suicide cannot be rational because there is no future in which to reap the fruits of such rationality. With my two sets of neighbours, on the other hand, the future remains wide open; my reactions to both sets is informed by my awareness that all of us are in the *middle* of our lives: the very way I bid farewell to the house-selling neighbours is informed by the fact that I could, if ever I wanted, look them up again, and be looked up again myself. They could even repurchase their old house back again. For in principle there is always 'plenty of time' in which to make amends, apologise, carry out vengeance, go broke or have 10 children. Saying goodbye to Ken Harrison, on the other hand, would be very different. Let us assume Harrison's request is granted. His decision, however rational, does not provoke anything like the responses to my neighbours. There is no admiration or envy, for he is a man quite literally without a future.

But if I am right about this, is it not a more general point about our last conversations with the terminally ill? In part, perhaps. But there is a very real difference in the fact that a terminally ill patient is *struck down* by capricious fate; were it not for the illness, he would very much like to be alive, and for as long as possible, just like the rest of us – that is the very source of his regret. There is still unease in the response to the terminally ill, but I suggest that, interestingly, there is no horror. There is unease born from the advent of grief in response to a beloved being taken away, to an awkward situation where it is not clear what to say, and perhaps to a renewed awareness of one's own mortality; but this is not the same as the response to someone *leaving* the world. Compare the difference between my wife dying and my wife divorcing me and moving away. The former is certainly traumatic, but there is perhaps comfort in the knowledge that my wife is going to 'a better place', a place where she can wait for me. When my wife is not taken away but takes herself away, on the other hand, this is an expression of will, of preference, of defiance, an unambiguous statement that I and my world are no longer good enough for her, and that our special union has turned out to be no more than a failed experiment. Certainly I may long for her return, but even if she does return I will know that she 'has it in her' to leave again. It is this defiant attitude, in this case irreversible, that is the root of the horrified response to suicide.

Brandt might well accept that we do respond with horror to the news of someone's suicide, but would say that this is merely a contingent response; it still makes sense to ask whether the act was itself rational or not, regardless of the responses to it. I reject this putative contingency and claim instead that the horrified response is essential to the meaning of the act of suicide within our language and culture. So much of our language and culture, so many of the routine assessments we make every day about the degree of rationality of ordinary acts and decisions, presupposes an open future.

Horror was the first element that Brandt's account neglected, and pity is the second. As before, I claim that pity is not an accidental response, but is essential to our understanding of the full meaning of suicide. Pity is more than sadness. One can be sad about a lot of things, from the end of summer to one's football team losing. Pity is a response of sadness to another's suffering, true, but crucially it involves some understanding of the sufferer's point of view, an understanding of what it is like for this particular person to suffer in these circumstances. In discussing the victim's rationality, Brandt is talking *about* the victim, as it were behind his back, rather than *to* or *with* the victim. Brandt maintains a clinical detachment, presumably in order to make what he considers a more objective assessment. Once again, I am not claiming that Brandt is incorrect in his discussion, but that he is missing the main point because he does not make much of an effort to understand the perpetrator's particular point of view on his own situation; instead, Brandt is looking for what it would be rational for *any* agent to do in the perpetrator's *type* of situation.

Why does the agent's point of view matter so much? Surely his behaviour, his situation, his reasons, as captured in a third-personal description using general concepts, is enough. Normally it would be, but that is because there is an implicit assumption in using the concept of rationality that the putatively rational person will himself be striving, on the basis of his beliefs (which may of course be mistaken) for the most rational outcome (the protection and promotion of his needs and interests), and so the two uses – the first-personal and the third-personal – coincide. If I tell my neighbour that the most rational thing to do would be to sell their house now, then this will support their own judgement to the same conclusion. There is usually one most rational thing (or several equally rational things) to do in a given situation by *anyone* who finds themselves in that situation, and you and I can, in principle, converge upon it, and help one another to discover it. Even if my neighbours' house-selling decision goes disastrously wrong (and the house prices rise just after they sell), I can still reassure them that they did the most procedurally rational thing given the evidence available at the time (i.e. real estate market trends). To summarise: it is normally *reassuring* to be told that one's behaviour is rational, and one is grateful to be told.

In contrast, imagine the following situation: Harrison has been granted permission to commit suicide, and now he hesitates. Brandt agrees with Harrison's earlier judgement that suicide would in fact be the rational thing to do. So what happens now? Does Brandt try to remind Harrison of how rational the act would be? Does he egg him on, tell him to grit his teeth and get it over with? Would Brandt tell him that he would do the same in his shoes?

It is no accident that this starts to sound like criminal bullying, and that suicide is perhaps the *only* thing that one person cannot advise another to commit. Would Brandt try to reassure Harrison's grieving parents by 'reminding' them that their son's suicide was rational, that everything had turned out for the best? And if so, should they feel, what – relieved? True, it is not clear what sort of response is appropriate to either Harrison or his grieving parents, but it surely would have to be one informed by pity rather than by a concern for rationality. There is no 'most rational' or 'best' outcome in *Whose Life?*; there is an awful situation resolved in an awful way. However strong Harrison's arguments for committing suicide, they cannot be compelling in the larger scheme of things, for there are still plenty of other reasons for Harrison, and indeed for anyone, to go on living. That he chooses not to means that he himself can see no other option, even though we can: and this is to be pitied. This difficulty in considering suicide to be the right or best answer is the second kernel of truth in the Absolutist position.

David Vellemean (1992) made a similar point in his argument against the legalisation of euthanasia. If voluntary euthanasia were a legal option for a patient, it would not be an option that could just be put on the table during a putatively non-directive counselling session. In the same way that I argued that abortion is not a neutral option, the mere existence of the option of legalised suicide implies a situation where it could be recommended, and so we would end up with the absurdity described in the previous paragraph. Suicide is an option that has to be contemplated and chosen entirely by oneself; there can be no higher authorisation, not even reason.

My point here is more than merely to distinguish between the metaphysical status of the act and its pragmatic significance in human relations (i.e. the sort of things one would in fact be inclined to say about the act or to say to the victim); there is no reason to think that Brandt would not acknowledge common norms of tact and decency. My point is somewhat stronger. With a rationally susceptible act like selling a house, the importance of the rationality or irrationality of the act is revealed by my legitimate use of the term in efforts to reassure or console. The fact that the concept of rationality 'runs out' (i) when I consider my own horrified reactions to the news of someone's suicide, and (ii) when I consider the sort of things that I am inclined to say to the victim or his family about the act means that the concept was not important in the first place and cannot do the philosophical work expected of it.

To clarify, there are two things I am *not* saying. First, that the concept of rationality is inapplicable to suicide *tout court*; many of us can imagine ourselves making a similar decision to Harrison if we ended up in his situation. Second, in claiming that the concept of rationality does not get us very far in trying to understand another's suicide, I am not thereby implicitly calling the act irrational: both concepts, of rationality *and* irrationality, run out when trying to make sense of one's own reactions and of the things one feels inclined to say.

Let us look at it from the other side of the coin, so to speak, by taking another suicide that I assume Brandt would regard as irrational: that of Michael Furey, the character from James Joyce's short story 'The Dead' (Joyce 1996). We learn of his sad fate from the story's heroine, Gretta, many years later. The teenage liaison between them was to end because of Gretta's family moving away. On the night before the departure, Furey appears behind Gretta's house to say goodbye. Already quite ill from some unknown but lethal ailment (probably tuberculosis), he was out on a winter's night without a coat. Despite Gretta's imploring him to return home, he declared that he no longer wished to live. Several days later she receives word that he died.

Despite Joyce's moving account, it is possible to view Furey with some impatience. Unlike Harrison there is a very real prospect of his life improving if he can just get over this phase – most of us have known something of what he is going through. So if we consider ourselves in a position to offer advice, we might remind him that Gretta will probably return one day, that they can correspond until then or that there are other fish in the sea, that he should count his blessings, etc.; or we might just shout at him to 'snap out of it', or to stop being so self-indulgent and to finish his supper. All these approaches assume that this is an irrational phase to be endured. But what does it *achieve*, philosophically, for Brandt to call Furey's wish, and then his act, and indeed to call him, irrational? As before, we have to look at the sort of contexts where there is a point to calling someone irrational, and those contexts will involve blame of some sort and an expectation that the person will himself see the error of his ways *in order to profit in the future*. 'Listen, don't buy the car just because you liked the salesman. It would be more rational to do some homework, to talk to my mate Hymie who knows about cars.' Here I am inviting the other to see the situation as I do, to see how my recommendation would better serve his needs and interests, or better satisfy his desires. I am also hoping to bring him to a position whence he can later look back on his past self and agree that he was being irrational: 'what got into me? Why didn't I see that the salesman was too smooth a talker.' To sum up: the concept of irrationality can always be unpacked in terms of a threat (or at least a risk of suboptimal outcome) to the agent's needs and interests, or at the very least in terms of counterproductivity in the agent's search to satisfy a specific desire. If Furey is dead, then calling his act irrational comes too late, and amounts to little more than name-calling.

So once again there is an essential temporality to our use of the concept of rationality and irrationality. When talking to Furey before his death, it is revealing that we do not stop at saying 'don't be irrational' once we see how deadly is his earnest; instead, our immediate aim is surely to keep him alive by reminding him what he has to live *for*. But there is nothing *rational* in simply staying alive for these things. There is nothing rational in longing for Gretta's return, in contemplating other fish in the sea, in trying to forget Gretta by absorbing oneself in work; it's simply what many of us do, what many of us find pleasure and meaning in. If he doesn't find pleasure and meaning in them, then in a very brutal sense those are the limits of his world. We can try to keep him alive by telling him that we were once in his situation, but why on earth should he care? After all, he will say scornfully, *we* don't understand the depth of his passion for Gretta, *we* have never felt quite like that with anyone else (Gabriel, Gretta's husband, admits as much). All our efforts are so meagerly hit-and-miss that it would be ludicrous to describe them as bringing someone to a more rational view of the situation, even if he does end up changing his mind; all that is then left is pity. In its charming Enlightenment optimism, analytic philosophy has never really been able to make sense of despair. Here is the sort of thing Brandt would say about Furey:

> Depression, like any severe emotional experience, tends to primitivize one's intellectual processes. It restricts the range of one's survey of the possibilities. One thing that a rational person will do is compare the [future] world-course containing his suicide with his best alternative. But his best alternative is precisely a possibility he may overlook if, in a depressed mood, he thinks only of how badly off he is and does not contemplate plans of action which he has not at all considered.
>
> (Brandt 1975 p. 380)

The problem again is that Brandt is still talking about Furey, rather than trying imaginatively to enter his perspective of the situation. The concept of depression seems to offer a causal explanation for misperception, but it is merely an empty redescription that says more about Brandt's frustration than Furey's perception. The assumption is that the 'best alternative' is *there*, in the situation; Brandt can see it, but why can't Furey? Is it a matter of drawing Furey's attention to it, and the 'bestness' of it will draw Furey's interest and override his despair? It is distinctive of my approach to ethics that I am interested above all in what Brandt could say to Furey in exactly this sort of situation. With the house buyers, I can say what I think, for we behold the same problem. But Furey's problem is his alone, and it will not be solved until *he* solves it. The mainstream conception of ethics involves the assumption that a singular set of reasons are there, in the situation, for anybody to behold and then invoke as part of a moral justification for action. Instead, the crucial point, is that reasons should be taken as expressing the way the reason-giver experiences the meaning of the situation. If Furey cannot see a reason to go on living, then it is naïve to think that another person might somehow be better placed or better equipped to see the reasons in

Furey's situation and to point them out to him; and it will be entirely inappropriate to wring one's hands and complain that Furey just 'wouldn't listen to reason'. Or rather, there is of course a possibility that an attempt (if not the first, then the second, or the third ...) to dissuade Furey will succeed, but this will not be on the basis of presenting him with logically compelling reasons; and it can only ever be a temporary success. Once someone like Furey has tasted the distinctively radical freedom associated with the serious intention to commit suicide, once he has discovered the distinctive peace from moving beyond the shared space of reason, then the option can never be ruled out again – as every member of Furey's family will find to their cost and distress.

Diane Pretty

In this section I want to consider a very special case of attempted suicide, or rather, of an explicit request for assistance with one's own suicide: the English legal case of Diane Pretty. This will bring a further dimension to our discussion of suicide because of the proposed involvement of a significant other, her husband Brian. It will also continue to reveal the limitations with invoking considerations of rationality or justification within the framework presupposed by mainstream medical ethics.

Diane Pretty was in her early 40s when she was diagnosed with motor neurone disease, in November 1999. Her condition then deteriorated rapidly, with the certain prospect of increasing paralysis affecting her whole body, loss of speech, and the need for tube feeding. By early 2001, she had decided that she no longer wanted to live, but by then was physically incapable of killing herself neatly, and did not want to face the slow, painful and undignified method of refusing to eat or drink. Her husband Brian offered to give her a sufficient quantity of barbiturates to kill her; i.e. he would put the pills directly into her mouth for her to swallow. Under section 2 §1 of the Suicide Act 1961 in English law, it is not a crime to commit suicide, but it is to assist someone else's suicide. On 27 July 2001, Mrs Pretty's solicitor wrote to the Director of Public Prosecutions (DPP) asking for assurance that her husband would not be prosecuted for helping his wife to commit suicide. The DPP refused. She then appealed the decision several times, and the case famously culminated with the European Court of Human Rights rejecting her appeal on 29 April 2002. Thirteen days later Diane Pretty died from breathing complications, under sedation at a hospice.[2]

In some respects, this case adds nothing new to the long-standing debate surrounding physician-assisted suicide (PAS). Mrs Pretty could well have asked a doctor to provide her with a lethal injection. As such, Peter Singer's (2002) typically utilitarian response to the Pretty case starts with the familiar rejection of the acts–omissions distinction: if the foreseen result is the same, then the ethical permissibility of both act and omission must be the same. In English law, any competent patient can refuse any treatment, for any reason, even when such a refusal will certainly and knowingly result in the patient's death. Switching off a ventilator in response to such a patient's request is legal. The only difference with the case of Diane Pretty is that she was not dependent on any such machine to survive – the only 'treatment' she was in a position to refuse was nutrition and hydration. Singer argued that competently requesting that a ventilator be switched off is tantamount to requesting assistance with suicide. Whether a machine is switched off or a lethal drug administered, the certain consequences are known and freely chosen by the patient. The law is therefore inconsistent, and the Prettys were trying to have the law changed.

The responses to Singer are equally familiar. Perhaps the most popular runs: there *is* a distinction between acts and omissions, between killing and letting die, for otherwise I would be morally responsible for any of the world's suffering that I could have averted through, for example, a donation to charity that I did not make. What both Singer and his opponent accept, however, is the importance – and possibility – of justifying their decision. I shall not enter this debate on either side, for I am again interested in the personal perspective. Singer elides the public and the personal questions, and seeks a single, universal kind of justification. However, it is possible to speak about the public debate being resolved by the rejection of the appeals, but this still leaves the personal dilemma of whether Brian Pretty should offer the pills to his wife anyway and accept the legal consequences: the possibility of a conviction of murder and a gaol term.[3]

A first view of Brian's dilemma would borrow from the discussion in Part I. The objective dilemma as described by, say, Peter Singer would not be the dilemma faced by Brian for the simple reasons that it is *Brian's* wife we are talking about; it is *Brian* who has to make the decision and *Brian* who has to bear the consequences of the decision either way. Singer may well offer advice, but it will be very much up to Brian whether to adopt that advice; and in adopting the advice he will *not* be acknowledging that Singer was 'right' (and that he, Brian, is now doing the 'right' thing) in the way that he might ask for Singer's advice about the best route to the Princeton cafeteria. Brian's good friends might be able to offer better advice, but Brian is still utterly alone with the decision of whether to accept the advice, that is, to adopt it and make it his. Because the observers are not implicated, there is a deep sense in which they cannot discover what Brian ought to do, but can only wait to see what he *does* do.

However, putting the dilemma squarely on Brian's shoulders is also misleading, for it assumes the horizontal atomism presupposed in the mainstream account, that is, that there is no essential link between Brian and other people. In reality, he is engaged in a continuous dialogue with Diane about what *they* should do. He is not making a decision about 'his wife' in a way that 'one' might make a decision about 'one's wife'; he is making a decision with Diane about Diane, the person with whom he has lived for many years. To put it simply, Brian is not interested in other people's wives, or in the responsibilities of their husbands to those wives. He is only interested in Diane, this person here and now, her suffering and her dignity. The goal driving his deliberations is not rightness or justification or rights or responsibilities, but 'Diane' and his particular understanding of what is best for her or of what she deserves.

In the end, Brian did not help his wife to die. What does his reluctance say about his *love* for Diane? To an observer, it could go either way. Perhaps he loved her so much that he was unable to help her kill herself, even if she asked him to, and even if he felt it was 'for the best'. This might be a special form of weakness of will, but surely an admirable one. (In the same way, Brian's close friends might not be able to bring themselves to recommend killing her, even if *they* sincerely thought it would be for the best; and Brian might well admire them for that.) On the other hand, if he did end up helping her, perhaps this revealed that his love for her and his desire to see an end to her suffering was sufficient to risk gaol for. And this too might be admirable.

Love is a tricky concept, however. Imagine that in the end Brian helps his wife to die. As an observer, we can say that the act was an expression of his love. But this does not mean that the concept of love enters his *experience* of the decision to act or not act: his thoughts are about Diane, her suffering and dignity, and the likely consequence of the prison sentence. Here again the traditional conception of ethical deliberation errs in taking the personal perspective to be on par with the spectatorial. Very often I can make the translation from the third person to the first person quite legitimately; if I see you in a shop, looking at the word 'discount' above a pair of jeans, I can attribute the following explanatory reason for your subsequent purchase: 'he bought the jeans because he liked them, and they were on sale.' And this could be the very reason that *you* would give, in the first person, for performing the action. But not all third-personal reasons are so translatable, a point well made by Phillips:

> Is not the form of the imperative 'You ought to heed these considerations if you care or if you are interested'? This is not so. To think otherwise is to confuse the conditions under which a man has reasons for paying attention to moral considerations with his paying attention. He will not have such reasons unless he cares, but the fact that he cares is not his reasons for caring. If, hurrying to the cinema, I stop to help the victim of an epileptic fit, while it is true that, in the absence of considerations of personal advantage, I should not have stopped unless I cared, it does not follow that my reason for helping him is because I care. The reason is to be found in the suffering of the epileptic.
>
> (Phillips 1992 p. 133)

Any observer can reach for underlying notions of interpersonal sympathy, and say that Phillips helped the epileptic *because* he cared. But this underdescribes the phenomenology of acting out of concern, for it is not the reason that *Phillips* would give in this case; it *could* be the reason, of course, but this would reveal an entirely different meaning to his actions. For example, it would suggest he was trying to impress someone who thought he was heartless. So the reasons that Phillips gives to support a judgement also reveal the full meaning of that judgement. To put it the other way around, Phillips's care is revealed in his ability to help for certain kinds of reasons that do not invoke the concept of care.

It is worth saying more about a neglected function of reasons in ethical discourse, and here I am summarising a more detailed argument made by Michael Weston (1975), among others. Recall the example in Chapter 2, of the doctor who decided to stay at home to look after his sick child, despite the hospital manager's belief that this was not a sufficiently 'good reason' to remain at home (where the expression 'good reason' was cited in the employment contract). It is the *agent's* reasons (i.e. the reasons he *gives*), therefore, once known, that make the agent's choice intelligible from the point of view of the agent. Reasons should not be thought of as 'out there', generated by the situation, ready to be discovered by the ethical enquirer seeking to justify his action. Instead, a reason expresses what is salient to the agent during his deliberations, i.e. how he sees the world. For the doctor, his child's illness was salient in a way that it was not to the manager. For Phillips, the epileptic's suffering was salient in a way that it was not for those who 'passed by on the other side' (to quote from the parable of the Good Samaritan). That an agent should know his reason for action is part of what we mean by the term 'reason for action', for our interest in reasons is precisely an interest in the agent's own account. To ask how someone knows his actions or intentions is like asking him if he knows what he is thinking – and the answer is not about special access to what he is thinking (for if 'special', this implies relative to someone else's access, which is non-sense), but rather that he *says* what he is thinking; the sincere statement is sufficient to establish that he *was* thinking such-and-such (absent good reasons to disbelieve him). The reason thus fulfils its function not by reporting something that was discovered, but by making the action intelligible. This attribution of an intention or reason goes together with understanding the action of another person as voluntary, and as such it characterises our relation to other human beings.

If Brian had helped Diane to die, and had then been asked by a reporter why he did it, it is hard to imagine him saying anything that would satisfy the McMahons of this world, even if he had been willing to defend his action in the context of an academic debate. Rather, the reasons that he would give would function solely in providing the contours of his experience of deliberation before the action.

Now consider an observer who learns that Brian Pretty had just helped his wife to die, and the observer says 'I can understand it, but I cannot condone it'. Under the mainstream account of justification, this can only be paradoxical. If you can understand it, that means that you accept Brian's reasons as ethically sufficient, and therefore you are committed to condoning it. In the 'perspectival' account I have been defending, there is room for an observer to condemn Brian's action unconditionally as murder, and yet to make the imaginative leap to understanding something of why he did it. The paradoxical double reaction is what is often expected from readers of fictional literature. Macbeth is certainly a murderous tyrant; but if that is all there is to him, then the play would have little to offer beyond the cartoon villainy of a fairy tale. 'Understanding' does not imply that I, the observer, would necessarily do the same in his situation, nor does it imply that the action was objectively right. Instead, to understand another's dilemma and plight is to be led to a certain humility in appreciating the vast ethical complexity of the awful situation in which someone like Brian finds himself; and to be led to accepting that I cannot say for sure how I would behave until I face something similarly awful myself.

The ethics of palliative care[4]

The thing that Diane Pretty most feared eventually came to pass. Not only did she have to wait to die, rather than choosing her own time, but the end itself was one of slow asphyxiation. Even if the discomfort was reduced by sedation, it was arguably not the most dignified way to go. And many people support euthanasia precisely for these two reasons: death is bad enough, but to lose control over timing and comfort makes it much worse. For although sedation reduces discomfort, it also reduces lucidity, so that one is effectively drugged to death, and the precise moment of death is difficult to specify if the person has been half-conscious for a while beforehand. In terms of the timing, there is a plausible fear of dying too soon, and therefore being unable to put one's affairs in order and say goodbye properly to all the people one was planning to; or too late, and therefore lingering like an unwanted guest after all the grand gestures have been made and the small talk has run out. In this section I want to examine these issues. If we are to discourage the Diane Prettys of this world from killing themselves, so the argument runs, then we have to be able to offer good palliative care.

The status of palliative care in medicine has always struck me as curious. Different people get different diseases and problems and so require different medical specialists, fine. *But we all die.* And so one would assume that good palliative care would be in a prime position in the medical package available to us. And yet if there is one thing that palliative care specialists complain about in unison, it is the lack of allocated resources, research interest, prestige and career ambitions among medical students. An obvious explanation for this is that medicine is all about curing, and it is all about technology, and palliative care involves neither; instead, it amounts to 'giving up', and there seems to be nothing more for a doctor – as opposed to other healthcare professionals – to do. So a first problem for palliative care doctors is that of justifying the allocation of more resources to their department and away from others. In QALY terms this would be very difficult. Well-funded palliative care will certainly improve present quality of life, but most of the patients will not even have a full year to be quality adjusted.

Answering this question will invoke some of my discussion in Chapter 6, of what it means for a life to have a shape. Before I do that, however, I want to add one more example of a suicide to the ones we have been discussing so far, and that is the character of Enobarbus in Shakespeare's *Antony and Cleopatra*. Enobarbus was Marc Antony's loyal adjutant of many years. However, during the period covered by the play, he becomes more and more concerned by Antony's preoccupation with Egypt and his distraction from the requirements of political and military office. By Act IV, Antony has engaged Octavius Caesar in a disastrously misjudged battle at Actium, and Enobarbus finally resolves to abandon Antony and offer his services to Caesar. On hearing of this defection, Antony is unhappy but not resentful, and poignantly orders all of Enobarbus's valuable possessions to be sent on to him. Upon receiving these, Enobarbus is astounded by such generosity of spirit, and filled with suicidal remorse. He resolves to 'find a ditch wherein to die'.

There are two distinctive aspects of Enobarbus's situation that are relevant to a discussion of palliative care: his inconsolable remorse and his desolate solitude. The question then is whether palliative care can offer anything to people in this position. Enobarbus's planned suicide is distinct because it is not a result of sheer bad luck, as it was with Diane Pretty and Ken Harrison, and to a lesser extent with Michael Furey; rather it is a result of his remorseful realisation of what he has freely *done*. Certainly Enobarbus was unlucky to find himself on the losing side in this particular conflict, but it was he who chose to abandon his master for reasons of pure self-interest.

Note that it is incorrect to say that Enobarbus commits suicide 'as a result of' what he has done, since this implies some sort of psychological causation between two contingent events. Instead, the suicide is the necessary culmination of the full awareness of the meaning of what he has done. It may be objected that he has not done anything *that* bad: after all, Antony had lost the battles against Caesar anyway; and Enobarbus did not betray Antony, only abandoned him. And by that point Antony has committed a series of political and military blunders unworthy of the great general, and so it could be said that he did not merit such loyalty anyway. But of course this is to adopt too external a perspective, for it is Enobarbus's perspective that matters most to Enobarbus, and his remorse is certainly intelligible as the sort of thing that *some* of us might experience in this situation.

So the phenomenology of Enobarbus's craving for death is different from the other three examples. Harrison and Pretty find life intolerable because of new physical impediments; without those impediments they would certainly wish to go on living. And neither Harrison nor Pretty would deny that others in their situation might well prefer to go on living, but they themselves simply find the impediments too great. They may curse their unlucky fate, and say to themselves 'I did my best, but it's not to be'. Enobarbus's remorse, on the other hand, comprises a devastating final chapter to the story of his life. This one recent act has effectively contaminated the whole, and silenced all other reasons to live: he has become worthless, despite the justifiable pride he might have had, up until that moment, in his accomplishments and relationships. And even if Enobarbus survived in an afterworld, his remorseful torment would continue. He cannot blame anyone else, much less fate, for his present misery, for misery is exactly what he feels he deserves.

While Enobarbus differs from Furey on account of the latter's innocence, the two have the same desolate solitude. Harrison and Pretty, we may assume, can perhaps look forward to death surrounded by loving friends and family. Enobarbus and Furey feel utterly alone, even if they were to be in a crowd. However, the key difference remains that Furey would recover instantly and entirely if Gretta returned (or, to be more cynical, if someone more attractive turned up). Enobarbus effectively has a death sentence which he cannot avoid since he cannot turn back the clock. If Antony were to offer to take him back into service and forgive him, this would make things even worse; for insofar as he conceives himself as guilty, he craves punishment, not forgiveness. And the most appropriate form of punishment is not just death, but an ignominious death, in a ditch, alone.

There is something particularly terrible about dying alone. By 'alone', I do not really mean alone in a room or a building. If a person takes a long time to die, so that family members have to take turns by the bedside, the last breath may take place when there is nobody present; but the family is nevertheless there throughout. Full solitude in death occurs when nobody knows that you are dying, because you do not want them to know. This terribleness might seem peculiar; for it is terrible even if nobody else can be distressed about the death, and even if the death itself is relatively painless.

If the full ignominy of Enobarbus's chosen death is difficult to grasp for modern ears, consider the death of a teenage drug addict, choking on his own vomit while shooting up in a dirty public lavatory. This goes beyond the problem of treating pain and discomfort and solitude, beyond the problem of writing the last chapter. Once again, it is tempting to dismiss the problem of dignity, since death is death, wherever it occurs, and the dead person will certainly not care after he is gone. But this is to ignore what we as a society have *made* of death, how it fits into our lives, and how therefore it must fit into my life. 'Made' is Cora Diamond's term, which she used in reference to animals: we cannot stipulate what animals are without first discussing what we as a society have made of them, what role they play in our lives. In the same way, we have to take death seriously not only as a personal event but also as a social one.

This question of dignity is no more apparent than in the mass dying facilities called nursing homes. There is an important difference between hospitals and long-term care facilities, of course: hospitals are only temporary accommodation, and that's where they fix people, or at least stabilise them. But the transfer to a nursing home is the beginning of death. I surely do not have to describe just how awful these places are, despite all the noble attempts to beautify them: the décor, the social activities, the private rooms, the pets, the variety of the menu. None of it can deny the environment of death: people come here to die, and they generally have a miserable time while they wait for it (with the exception of the dementedly cheerful, on which, see the next chapter). Insofar as a middle-aged person thinks about it at all, it is only to resolve to themselves that they will avoid such places like the plague when it comes to their time. And yet when it comes to our time, we are swept up in the current of decline and care and no longer have the strength to protest, and no longer have viable alternatives to fight for once we cannot look after ourselves. Far better, we may say, to go out in the fire of some passion, some cause, than to wither dismally in these antechambers among the company of walking corpses. As I hope I have made clear by now, death itself need not be a bad thing, it can be embraced, one can reconcile oneself with it, but that becomes difficult once one has lost the capacity to care for oneself and ends up in one of these places. The system of nursing care in Western society remains the last great taboo.

In this roundabout way, then, I would justify an increase of funding to palliative care. The more general point flows naturally from my discussion of the last chapter in the story. Dying people need time and space to finish the last chapter. As a result, palliative care is not only about treating this person here and now for their physical pain; rather, it is about treating this person's *whole life*. We could even pursue the argument in crude 'retrospective QALY' terms. If the person is allowed to wrap up their lives properly, then the aggregate quality of their whole life improves, whatever the varying quality of the different episodes as experienced at the time. That amounts to very many life-years 'added'. In the case of Enobarbus, the remorseful man needs someone to talk to in order to articulate his remorse and to understand it, even if it is to culminate in his death. He needs to understand the place of his remorse in his life, even if this makes him more unhappy. For many, lucidity is more important than comfort in one's final moments. And all this needs another person to talk to, to tell one's story to, and to leave behind with one's story.

This is perhaps easier said than done, and requires special training and a rich life experience. But dying people are also more difficult to talk to because they are dying – and this challenges a background assumption about the other people whom we encounter and engage in conversation with. Let me recapitulate here one of the central themes of this book, relevant to our understanding of those both entering and leaving the world.

The paradigm of trying to understand another person is my ordinary encounter with healthy people, those with an open future. And yet even here I have problems understanding; often I lack sufficient interest or patience to even begin to try. Close friends and spouses can sometimes behave in utterly inexplicable ways. And yet throughout, there is the assumption that *if* I had enough time and patience, *if* the other wanted to explain things, then I would come to understand 'what you're going through', or 'what it's like to be you'. This assumption runs together with another one: there is 'plenty' of time left, with plenty of possibilities to explore. If I haven't learnt to play piano yet, there's always next year. If you and I cannot understand one another, there is always the possibility of a deeper friendship *from now on*.

With the dying patient, the unbridgeable fact of the encounter with the doctor is 'I'm dying, you're not; and our relationship will end very soon'. The healthy doctor and the relatives can look forward, can make plans, can learn from their past mistakes and apply such knowledge in the future; the healthy can take new evening classes, new hobbies, acquire new trinkets, plan for travels to new countries, encounters with new people, always moving forward relentlessly. As I put it in Chapter 3, when I say goodbye to someone, I can always see him again. Even if it's difficult in practical terms, with enough time and effort I can find them; they are still somewhere in the world. Even if I don't like them, and don't want to see them again, there is still the potential for reconciliation – which means I have the ongoing *choice* to refuse to see them.

All this changes with imminent and inevitable death (where 'imminent' refers to the subject's attitude, rather than to an event at a determinate moment in the future). It's now time to stop planning and to start winding down; to realise that there are certain projects I will never do, certain countries I will never see, certain acts I will never be able to make up for, certain people who will never forgive me. The doctor, and the visiting relatives, know this, and the dying patient knows they know this. They can both pretend that this is a normal encounter, they can collude in various recovery plans, try to protect each other from the truth, but at some point this will no longer be possible. Pretence and collusion is not necessarily a bad thing, and both sides can be forgiven for trying to drag out the pretence and collusion. Because after all, what is the alternative? What can we talk about if we are being completely honest? There is a clear advantage, not only psychosomatic, in the patient remaining focussed and in control and hopeful as long as possible. But there is then a risk of clinging to life in a way that is not only undignified, but may induce panic when the last treatment fails and death cannot be ignored any longer.

Nobody has described this process of dying better than Tolstoy in his 'Death of Ivan Illyich'. Illyich does not have the sharp remorse of Enobarbus, but he has a more general regret that he has allowed himself to be distracted in his life by things he now considers trivial, especially social prestige and career advancement. He has discovered that the so-called friendships he cultivated as part of his social and professional world could not endure his fall from grace, and he feels profoundly alone. The extent to which other people actually *do* care for him is unclear in the story; and perhaps he is unable to notice that because of his self-pity. But Tolstoy captures well the distinctive attitudes that he remembers being able to take towards actual or potential relationships during his life, relationships based on mutual affection or advantage, with the full knowledge of a range of possibilities for greater or less intimacy to choose from. Now that he is dying, on the other hand, he is trapped not only in the sick role, but also in the dying role. Others seem to behave towards him with pity or condescension; they seem to covet his job; his pallor and confusion ruins their parties; his lack of hair, his smell, and his wheelchair obstructs conversation. Slowly they begin to think to themselves 'I wish he'd get on with it'. Slowly he comes to hate the living: for surviving, for planning, for stepping over his still-warm corpse into their future.

How does one tell, from the inside, when hope is no longer appropriate? Indeed, how does one tell from the outside? This will be a question for both doctor and patient, and of course such questions do not apply only to the dying scenario; there are similar questions in any relationship. How can I tell – how can I be told – that it is time to give up, to accept defeat, in my marriage, say? How can I tell that it is time to give up on my drug addict son, not in the sense of physically abandoning him, but in the sense of switching into 'managing' him, and refraining from hope and emotional investment as far as possible? How can I tell that it is time to write off my friendship with Ursula, while preserving a veneer of affection, because we still have to get along at work, and because we can still be useful to one another?

And yet we are back at the paradox. We're all going to die, and we all know we are going to die. So none of this should be any surprise; there should be no denial or avoidance or hushed tones in discussions on the subject. After all, none of us are making plans for things in a 100 years' time. A fear of pain might well be rational, but surely not a fear of death, it's something we can plan for all our lives. Certainly there are some people who are less reflective about the shape of their lives, and who will not be interested in anything more than pain relief at the end. Certainly there are some who become morbidly reflective and needlessly depressed at the end. In either case, good palliative care should be able to offer a full range of assistance.

8
Making Sense of Dementia

In the last three chapters, I have been speaking almost entirely about the lives and deaths of competent people. It is now time to consider the incompetent. In this chapter, I want to bring together some of the above thoughts on lives and relationships into a central striking example, that of dementia. There is a famous case in the philosophical literature on dementia, that of Margo.[1] Margo is an old woman, and severely demented: she cannot speak properly or understand questions, she does not seem to recognise close family members, she is dependant on an assistant for all activities of daily life, and she does not seem to have any pleasures beyond the infantile. Most importantly of all, however, she seems to the researcher to be one of the happiest people he has ever met: she does not seem to show signs of depression or even of longer term frustration.

In the last chapter, I introduced two concepts that I claimed were more useful in trying to make sense of suicide: horror and pity. In this chapter, I will again discuss pity and ask what exactly we are doing, and whom we are pitying, when we pity the demented. And rather than horror, I will consider the emotion of fear, and ask what exactly are we afraid of when we claim to fear becoming demented. Of course some demented patients often become distressed and confused, and so it is less problematic to understand why one might fear such a condition and pity anyone who was in it. But for the purposes of this chapter, I will only be discussing Margo, assuming her to be content almost all the time. Pity and fear are still appropriate, I will argue, although in a different way.[2]

First, some organisational preliminaries, going beyond the original research. I will take Margo to be 70 years old, and will take her illness as having shown the first signs when she was 60. By that age she has had several decades of normal and fulfilling adult life: job and promotion, long-standing friends and family, hobbies, political opinions, musical tastes etc. She has had the chance to lead and shape her life according to her relatively stable conception of the good. She has always reflected on who she is, on where she has come from and on where she would like to go next in the short and long term. She has always been a proud woman, and proud of being in control of her life. Next, I will also assume that her husband is dead, and that her only immediate family member is her daughter, Alice, who is now 40 (i.e. born when Margo was 30) and single. Margo was living with Alice until recently, and they were relying more and more on part-time carers coming in, but in the end Alice could no longer cope and arranged for Margo to be transferred to a nursing home, where Alice now visits her regularly.

Although this chapter is about a clear case of incompetence, some of my conclusions will not apply to the congenitally incompetent, even though such a person might behave in the same way as Margo, and may well – as far as we can tell – experience the world in the same way. The difference is that Margo has had a full, competent life and *then* become demented, even if she seems unable to remember anything of it. It is the decline that interests me here, the loss of the full adult life, the transition to a mere shadow of Margo's former self. Nor will I be discussing other kinds of mental impairment, such as psychosis or depression, although the dementia might of course be associated with both.

What is the nature of Margo's dementia? Her prognosis at 60 of a slow and inevitable decline in higher intellectual function comes to pass. Early clinical symptoms included memory loss (especially of recent events), decreased ability to concentrate and solve problems, mild emotional instability (partly frustration from awareness of the decline). This progressed to disorientation and confusion, inability to recognise close friends and family members, inability to carry out daily activities and personal care, and a loss of any concern for personal care. Drugs can slow the decline, but not halt or reverse it. It is compatible with relatively good physical health, and so can easily last ten or more years after diagnosis; 25 years is not uncommon. Now and then traces of Margo will return, certain characteristic phrases or gestures, but their significance as attempted communication by Margo is weakened, because the phrases and gestures are isolated from any appropriate surroundings in her life. There does not seem to be a Margo 'behind' them.

Beyond this brief symptomatic description, I shall assume that the phenomenon of dementia is familiar enough, as is the fear and pity that I am talking about. John Bayley's (2002) biography of his wife

Iris Murdoch has been widely read, and was made into a film. Together with personal experience, these artworks provide the best source for philosophical intuitions.

The paltry philosophical literature on the subject only focusses on two interrelated problems: the first is not specifically about dementia, but about the sorts of conditions that might prompt the legalisation of non-voluntary euthanasia, especially as a result of an advance directive; the second is a question about the identity of the patient, that is, whether the demented patient before us now is the 'same' person as the patient of ten years ago. In the next section, I shall consider the nature of these problems, but only as an attempt to show that they miss the point. (I shall explore the problem of euthanasia in more detail in Chapter 10.) I will then explore the concepts of fear and pity, using some of the conclusions from previous chapters about what it means to lead a human life.

The problem of personal identity

Is the competent Margo at age 60 the *same* person as the demented Margo at age 70? In terms of the singular path traced by her ageing body through space and time, of course she is. But consider Derek Parfit's account in *Reasons and Persons* (1984). According to Parfit, two such individuals will only be the same person if there are sufficient 'psychological continuities' between them: roughly, if the person at 70 can remember enough of experiencing life *as* the person of 60, and if there are enough similarities of personality, interests, values etc. This means not just getting enough of the details right about what the 60-year-old saw and heard and did, but also having sufficient experiential memories of *being* that 60-year-old seeing and hearing and doing those things (and assuming that those memories are sufficiently accurate, of course). This account becomes plausible when we consider those cases of ageing Nazi war criminals who seem to have genuinely forgotten committing the crimes of which they are accused – assuming it is not an elaborate ruse to evade justice. If the war criminal has genuinely forgotten, then it is hard to punish them, in the full meaning of the word, when they cannot understand the deliberate infliction of suffering (e.g. a prison sentence) *as* punishment for something they actually did. After all, justice must not only be done, but must be seen to be done, and that includes being seen by the defendant as well.

Is this radical division of personhood plausible in the case of Margo? As far as we can tell, Margo does not remember anything of her earlier life. Her room is decorated with photographs of the earlier Margo, of people whom the earlier Margo knew and places where the earlier Margo had lived, but the present Margo evinces no clear behaviour that would suggest she recognises any of the photographs. More troublingly, Margo cannot recognise her own daughter, nor any of the other friends and family members who come to visit her. Although it is not clear what part of others' speech she understands, she does not demonstrate any recognition of names of people and places that ought to have special significance for her. Accordingly, in one clear and familiar sense Margo at 70 seems to be an entirely different person in every respect. The problem is made more acute if we imagine that Margo at 60 signed an advance directive, stipulating that 'she' was not to be given life-saving treatment if she were to become severely demented. If Margo-at-60 and Margo-at-70 are indeed two different persons, then it seems problematic to condemn the latter to death from an easily treatable infection. It is even more problematic when we consider how much Margo-at-70 seems to be enjoying her life, and how little pain, physical or psychological, she seems to be suffering. While her enjoyment is quite infantile, this still suggests that she should be offered at the very least all the legal protections that infants are offered, and that any advance directives written by 'other' people should simply be ignored.

The 'advance directive problem' was first formulated explicitly by Buchanan (1988). In attempting to solve the advance directive problem, Ronald Dworkin introduces a distinction between a person's 'experiential' and 'critical' interests. A person's experiential interests have to do with his enjoyment of those particular experiences. A person's critical interests, on the other hand, are

> interests that it does make their life genuinely better to satisfy, interests they would be mistaken, and genuinely worse off, if they did not recognise. [...] I do think that my life would have been worse had I never understood the importance of being close to my children, for example, if I had not suffered pain at estrangements from them. Having a close relationship with my children is not important just because I happen to want the experience; on the contrary, I believe a life without wanting it would be a much worse one.
>
> (Dworkin 1993 pp. 201–2)

Dworkin is careful not to be prescriptive about the *content* of the critical interests: he can accept that others might not consider a close relationship to their children to be so important. But each of us will acknowledge some distinction between the critical and the experiential. Dworkin's next move is to say that critical interests are typically much more permanent through the course of one's life; as such they are much more constitutive of who one is, whereas experiential interests may change radically from one decade to another. The fact that Dworkin enjoys watching televised football is not so important to who Dworkin is; and if Dworkin were no longer able to watch televised football this could not be considered a great loss. But if Dworkin lost touch with his children, and more importantly, was not particularly disturbed by having lost touch with his children, he would have lost something *even if* his later self would not be suffering – in an experiential sense – without his children in his life.

These considerations can now be applied to Margo. On the assumption that her daughter Alice was one of Margo's critical interests, Dworkin concludes that Margo-at-70, no longer able to recognise her own daughter, has lost a crucial critical interest, and has therefore become *less* of the true Margo. (Her experiential interests have changed radically, and she seems to be enjoying her new life, but this in itself does not compromise the claim that she is still Margo.) If we bring the question back to the advance directive that Margo signed at 60, this should take precedence, Dworkin concludes, because it was written by the real Margo. This is interesting because Dworkin's reasoning goes in exactly the opposite direction as Parfit's.

I disagree with Dworkin's distinction, and I disagree with the conclusion that he draws from it. The critical/experiential distinction is valid enough at any particular time, although we would normally use the less technical vocabulary of 'caring' and 'liking'. And there is nothing particularly mysterious about Dworkin passing up the great game on the telly in order to help his son with his homework. But critical interests are surely not as permanent as he thinks they are, and the best example of this is children. Before I had children, I was driven by my career, by ambition, by greed, but then I softened and found that it didn't really matter if I didn't get the promotion since I wasn't prepared to put in the extra hours away from my new family. Or just the opposite pattern: before I had children, I had all sorts of noble sentiments about them, about the importance of the family, about all the things I'd do for them and with them; after I had them, I suddenly discovered how much work they are, how ungrateful they are, how they are not interested in the skills I want to foster in them. Indeed, I might even discover that I don't particularly like them. Since critical interests change over time, it is not clear *which* set of critical interests (i.e. those of which particular age) should be considered sufficiently dominant to define the person's true character.[3]

Imagine that Dworkin's children are happy to leave home at 18, and Dworkin is happy to see them go. Finally he is able to throw his considerable energy into the new book he's been planning. The years pass, and he finds he has less and less in common with his kids, their lives are diverging in a perfectly understandable way. If I were then to put it to Dworkin that he had lost one of his critical interests, surely he would reply that this interest pertained to an earlier Dworkin, not to the present. And yet Dworkin is not a different person for the change: in Parfit's terms, he certainly remembers enough of being Dworkin at a younger age, and he has certainly retained enough of his personality, interests and values. Which, then, is the 'real' Dworkin? Surely the later version, precisely because that is the version that we are encountering *now*.

The conclusion from the above is that people's critical interests can change without them changing identity. But there is a second point that I want to raise against Dworkin that is much more salient to the discussion of dementia. Dworkin's account (like Parfit's before it) is atomist in the sense first described in Chapter 6.

This means that he is trying to find the necessary and sufficient conditions for calling the earlier and later versions of a human being the same person. He believes that if one can look hard enough at the person, at their interests, at their life, then some link between the two versions will or will not be manifest. If such a link is not manifest, we can then say that the body has been usurped by someone else, in this case someone with none of the original critical interests. The problem, I will argue, is that when Dworkin considers the subject's relations to other people, it is only as revealing some further element about the subject. What he does not consider is how the relationships with other people (especially 'significant others'), and the other people's attitudes to the subject, might themselves be partly constitutive of the subject's identity.

To put it crudely but succinctly: the fact that other people *call* her Margo or Mum and *treat* her as Margo or Mum helps to *make* her Margo and Mum, the Margo and Mum *she has always been*. We see the sense of the claim that it is Margo in these reactions. 'Margo' is the name that almost everyone has been using to address her all her life, and it is what the nurses call her now – and not merely as a matter of convenience or politeness. For Margo to have a name implies at the very least that she has been formally named, and this implies the presence of parents in her life, of a naming ceremony, of a birth registration. Her parents continue to exist in Margo's very name, even if they are dead, even if Margo cannot remember them, even if she cannot recognise a picture of them. 'Mum' is what Alice has been calling her all her life, and it is *still* what she calls her, for Margo is still her mother, she is still the same person who gave birth to her, and of whose flesh she emerged. Alice does not need to read that name off the label on Margo's door, as some of the nurses do. There are pictures around Margo's room, and they are pictures of Margo; there are paintings painted by Margo. Margo may have changed radically in many aspects of her personality, but there is no doubt at all that it was *she* (the woman standing before us) who painted these paintings. Any attempt to 'split' Margo into a Margo at 60 and Margo at 70 does not understand the nature of a single human life, either as it is lived by the person, or as it is understood by that person and by others. Being with Margo, Alice may discuss shared experiences and stories from the past. The conversation is a bit one-sided, certainly, but it is nevertheless a conversation in the sense that Alice *attends* to her mother's reactions by, for example, reminding her 'what she always used to say about Dad'. Importantly, she attends to her mother in a way that is very different from how she would address a human infant, since with an infant there is no past experience to condition the present.

But if I am saying that Margo is the same person, does that mean that the condition stipulated in the advance directive has come to pass, and therefore that she should be denied life-saving treatment? Perhaps. But to answer this question we first have to understand what Margo-at-60 feared so much about becoming demented, even with the prospect of a life of infantile pleasure in an eternal present.

Fear, pity and mockery

In what follows I shall assume that it is intelligible to fear death, where death will be understood as annihilation without any afterlife. Margo at 60 is diagnosed with early-stage dementia, and all of a sudden she dreads the future. But what exactly is it she dreads? Imagine that you were offered a choice between a quick death, and a long period of severe dementia followed by death; I suggest that most of us would choose the former, but it is not immediately clear why. If in her demented state Margo remembers nothing of her past, has no conception of the future, cannot recognise her family and friends and has no higher order thoughts about the world, then it could surely be said that she is effectively dead. In which case we should fear the two outcomes equally.

However, because of Margo's continuing enjoyment of life, it would seem at least a little preferable to immediate death.

The most important difference is one of *authorship*. Within the limits of adversity and talent, I strive to be the author of my life, to write the stories of my past, and to plan the future with some view to the values that have shaped my life until now. This much was the insight of Dworkin's critical interests. When trapped by the bad guys on the hilltop, I can go out in style by careening off the cliff in my Aston Martin. My death then becomes another expression of me and of my life. Even when I am bed bound and wasting away with physical disease, I can rage, rage against the dying of the light, or face the inevitable with quiet dignity; there is still room for a choice of how to present myself to the world. Even when I lose control of my mouth or my limbs or my bowels, it is still *me* who is suffering the loss; I am still aware of the decline as a decline, and I am still the author of my last thoughts in the sense that I can seek to understand my predicament in terms of, say, some test of courage or some divine punishment. I am the author right to the end. Dementia is so different because my life continues without my authorship – that must be one source of our fear of dementia, that our lives are being taken away from us, while leaving some dim awareness. Indeed, the worst thing about moderate dementia may be the inability to hold carefully articulated suicidal plans in mind long enough to execute them.

Can we say that dementia is worse than falling into an irreversible coma? After all, there too there is a loss of authorship, as well as a loss of psychological connectedness with the earlier person. However, in a coma it is easier to see the body as either entirely unoccupied (in which case it would be dead), or as occupied by a sleeping Margo, one who could in principle wake up and be the same as before. In addition, the sleeping figure, like the dead body, can be much more dignified in appearance, merely resting after the long day's toil. After all, sleep is what Margo-at-60 did every night, only to rise again the next morning as her usual self. Whereas Margo has not behaved like an infant ... since infancy. This second infancy is anything but dignified: she drools, she wets herself, she giggles hysterically, she mutters nonsense, she stares wildly at nothing in particular. Is *this* all that Margo's life amounts to? Is this her last chapter?

Perhaps it is not fair to call this a *loss* of dignity, just because of a little drooling. After all, there is surely a sense whereby the congenitally handicapped, who may behave in much the same way as Margo, can nevertheless achieve a certain kind of dignity; it would be offensive to suggest otherwise. Margo should be valued and appreciated as the unique individual she has become. It is true that there would be no difference between the handicapped person and the 70-year-old Margo if we met both of them as strangers. We would quickly discover that our encounter was to be guided by slightly different rules, but no matter. We would speak more slowly, and more articulately – but this need not imply condescension at all. We would change the subject matter of the conversation, if indeed there is to be a conversation in words. But the contours of dignity are defined by the wider context of the life within which it is displayed and aspired to. The congenitally handicapped are capable of dignified and undignified behaviour, behaviour which is recognisable as such by carers, and understood as such by other members of the handicapped community, behaviour which they can be intelligibly praised and blamed for. But just as Parfit errs by 'cutting off' Margo from her past, because of the lack of psychological continuity, so too does the analogy with the congenitally handicapped break down when we look at the photos around Margo's room in the nursing home; Margo has several decades of a past as a competent adult. Just because Margo does not seem to remember her past, or remember the people with whom she had close relationships in the past, does not mean that the past has disappeared. As a stranger I can still discover it, as the nurse explains the photos to me and describes the special significance of the trinkets on the dresser, and as Alice recalls her youthful family life. What I am discovering is not an isolated story, but the biography of *this* person, Margo, that stretches out behind her like the wake of a ship. And I am also discovering what dignity has been lost, because I now understand better what Margo in this photograph, taken shortly after her diagnosis at age 60, dreaded so much.

Perhaps this is not quite right. If she cannot experience her lack of dignity, then what does it matter? As such it does not make sense to fear the lack of dignity in this context. Instead, it could be argued, she fears the process of losing her dignity, gradually, over several years between the age of 60 and 70. Her awareness is not of what she will become but of what she is becoming: her increasing loss of memories will at the very least frustrate her, embarrass her, but gradually frighten her as she disintegrates before her eyes. She fears the increasing threats that seem to present themselves in a world that is becoming less and less familiar. As such, Margo at 60 fears the anguish that comes with other forms of neurosis and delusion. This revised understanding of Margo's fear at 60 is mistaken, however, for the gradualness is not essential to dementia. Imagine if Margo-at-60 were again offered the choice between instant death at a certain moment in time (say, by a massive heart attack), and instant dementia, that is, without the progression to dementia and without the awareness of the gradual loss of one's memories. Both experiences would be identical from the first-personal point of view. And yet there is room for the thought that dementia is worse than death, and worth fearing more than death.

This is because it is possible to be harmed without being aware of it. A banal example might be where I am on holiday and burglars clear my house out. I am not harmed upon discovery of the theft; rather, I was harmed at the moment of the theft without being aware of the harm. A more controversial example is posthumous harm, where it makes sense to say that Hector was harmed when his dead body was mistreated by Achilles, even if Hector was unaware of it. In the example of dementia, what Margo fears is the harm that will be done to her and to her dignity by her future demented state. There is more to say about the special quality of the harm, however; in a moment I will be suggesting that it is akin to *mockery*. Before that I want to consider the complement of Margo's past fear, and that is Alice's present pity of Margo. It is as if both Margo-at-60 and Alice are spectators on Margo-at-70; the one feels fear, and the other pity, and both these concepts are linked.

Normally, we would pity someone for their clear suffering. But again we have the problem that the demented Margo does not seem to be suffering at all; quite the opposite. Does this mean that pity would be inappropriate? Perhaps the better concept to describe the feelings of Alice and the nurses is sadness, but even that might be inappropriate, since it is as if Margo could say 'thanks, but don't be sad for me, I'm having a great time!' Perhaps the better concept is one of grief for the lost Margo.

I am going to insist that it is pity. First, grief is not right for the simple reason that Margo isn't dead yet. One can only grieve if one takes Parfit's two-person conception seriously and sees the present Margo as someone else. But an essential part of the problem with dementia is that Margo has not been *allowed* to die; she is retained in this limbo until her body finally collapses, which could take many years. In addition, grief has a different phenomenology than pity; grief subsides with time. If my child dies, my grief will be heightened when I watch a home movie of her playing, but each time I watch the movie it may become less painful, or it may at least become a tolerable, bittersweet sort of pain. Whereas with Margo, on the other hand, Alice feels a sharp pity every time she visits her in the nursing home, precisely because there is not a death that can recede into the past. Finally, my knowledge that my child has died is compatible with some comfort of knowing that she has gone to a better place. But Margo remains stubbornly on this earth. Sadness is partly correct, but it is not enough to capture the full emotional response since it is not directed enough *at* Margo. Certainly we feel sad in response to Margo, but we can feel sad about plenty of things in the world, things that do not necessarily even involve human beings. In feeling sad about Margo, we are really feeling sad about Margo's situation. But the pity we feel is not for the situation, not for Margo's losses, but for Margo herself.

This is still not quite right, however, precisely because of the fact that Margo cannot appreciate her situation as pitiable. Sometimes I might pity someone who would find my pity condescending, who would deny that there is anything about his situation worth pitying. I might feel that he is simply mistaken, and hope that he might be brought to see his mistake, at the very least many years later. And yet it will be impossible to bring Margo to see, even in principle; she is beyond the 'reach' of pity. As a result, I want to suggest that it might not in fact be Margo as a *person* who is the object of our pity, but rather Margo as constituted by her *life*. What we are pitying is the consequences of the dementia on her life. In order to expand this notion, however, I need to bring in the concept of mockery.

At the very least, being struck by dementia is a misfortune, certainly. It is a misfortune that diminishes the person, makes him a shadow of his former self, certainly. But that does not seem enough. Here I might have to appeal to the reader's direct experience, or at least their imaginative engagement with John Bailey's attitude to the dementia that struck Iris Murdoch. The pity that Alice feels for Margo has a definite aspect of anger. The fear that Margo felt at 60 upon learning her diagnosis has a definite aspect of horror. This anger and horror is not the response to a merely unfortunate situation into which I was plunged while being fully aware of the situation and of myself in it. And it is not the more abstract kind of anger associated with indignation about injustice. Somehow it is more personal than that. I want to suggest that dementia's distinguishing feature is the fact that it makes a *mockery* of the person, of her life, and of her relationships.

Mockery might seem an odd choice in one respect. However much one might fear mockery, there is a sense in which one *ought not to* – 'sticks and stones may break my bones but names will never hurt me', after all. To engage in mockery is already base, and so the only person who comes off badly is the mocker, not the victim. If the mockery stings, it may be because of one's vanity: one should be comfortable and truthful enough with oneself that one can ignore the mockery, just as we are taught by our parents to do in the school playground. But mockery involves more than insult or derision; mockery can also involve a counterfeit or absurdly inadequate representation, in this case of Margo's life. And there is no author of the mockery to ignore, except perhaps fate itself.

It is this absurdly inadequate representation that undermines the value of the whole life. The dementia cannot be taken as a cruel aberration from an otherwise noble life, for that would be to espouse the two-person conception too closely. There is a sense in which Churchill's greatness was not undermined by the dementia of his last years, but that sense has more to do with the public portrait of Churchill's achievements since his death, a portrait that does not normally include reference to the dementia. For the biographer, just as for the nurse tending Churchill through his final illness, there is a life story spoiled by the ending.

There remains a problem. Surely there is a question about reality and appearance here: it is the appearance, i.e. the appearance *to me*, of the book that has been spoiled, not the book itself. The book itself is unchanged. Similarly, it is *my* understanding of Margo's life that is spoiled and mocked, not Margo's life itself. Let me call this the 'realist' objection.

There are three aspects of a response to the realist objection. First, any conception I have of Margo's life will be of the whole life, *up to now*. This means that my conception of Margo's life will be different when she was 60 than when she was 70 – but so be it; that is part of what it means to say that I know someone, and does not undermine any judgement I make about Margo in advance of the years she still has left to live. Second, any conception of Margo's life must be *somebody's* conception; it is incoherent to suggest that Margo's life has any meaning independent of all conceptions of it. I can summarise both these aspects by saying that the events of Margo's life do not have a fixed significance. The events themselves were fixed, this much I can grant the realist: there is a fact about whether Margo met her husband in Bognor Regis or in Lyme Regis, and the standpoint of the rememberer (Margo or anyone else) will not change the fact. But the significance will shift: was it a good thing or a bad thing that Margo met her husband? This will depend on the context provided by Margo's entire life up to and including the moment of observation.

The third aspect of the response to the realist objection is to say that anybody capable of making an *authoritative* observation of Margo's life, anybody in a position to appreciate the significance of a particular event within the context of her life, will have to know her quite well; and to know her quite well means to have been in a relationship with her, of sufficient depth and for a sufficient length of time. (The 'knowledge' that private detectives and biographers can have of their subject is quite different precisely because it is not reciprocated.) This means that the realist is mistaken to believe that there exists some neutral account of Margo's life, one that will be invulnerable to the mockery of the dementia in her final decade. The reality of Margo's life exists most in the lives of those who know her best, the ones with the close relationships who have watched her decline, the ones who will feel the sting of the mockery, even if Margo does not feel it herself.

The realist should not assume that the only alternative to his position is some rampant subjectivism, whereby whatever Alice remembers about Margo is true. After all, Alice's memories are answerable to the verifiable facts of Margo's life, and the determinate significance of the events remembered by Alice will be subject to the constraints of internal and external coherence (e.g. to norms of human behaviour, to other people's memories, etc.). But this much of the realist objection is correct. Alice is certainly concerned to *do justice* to Margo's life, and her disagreement with the nursing home management about what is best for Margo is structured by such a concern.

9
Human Bodies[1]

Part I involved a discussion of the medical world, and how extraordinary it is. In this chapter, I want to take this issue further by looking at just how extraordinary the attitude to human bodies is in that medical world. By looking at the meaning of bodies, alive and dead, I am again contrasting my account to the biomedical account of the body as studied by doctors. This investigation is not merely of esoteric interest. If I am right, then it will have certain profound implications for the communication between doctors and patients about treatment decisions, and between doctors and relatives about the bodies of the deceased. This latter point is especially important because of the recent scandal surrounding the covert retention of organs at the Alder Hey Hospital.

Every healthy human body has a gall bladder, and therefore, so do I. But I confess right away that I do not really know what a gall bladder is, and unlike the heart or lungs I had not heard of it until recently in my adult life. I don't know what it looks like, what it does, or how I would feel were it to malfunction. I know there's supposed to be one in my abdomen somewhere, I know it's an organ that probably secretes something, but that's about it. As far as I can recall, I've never talked to anybody about *my* gall bladder (as opposed to gall bladders in general); that is, I've never shown any symptoms that would have warranted an explanation from a doctor in terms of a malfunctioning (or missing) gall bladder. I've bumped into the word 'gall bladder' here and there, but have never been curious enough to enquire further.

But for all this, I *know* I have a gall bladder. And I certainly don't have any good reason to doubt that I have one, and a properly functioning one. But what is the nature of this knowledge? On the one hand, of course, it could be a straightforward combination of inductive knowledge (if I am a doctor, I have seen gall bladders in live and dead bodies) and knowledge by expert testimony, and my aim is not to challenge either of them as a reliable source of knowledge of the external world. On the other hand, there is something very special about being *in my body*. Certainly, my abdomen is a container like any other, and someone with the right skills and tools can open it up and inspect my gall bladder. In principle there is no reason why I couldn't do it myself, with a powerful local anaesthetic and a carefully positioned mirror. But it is important that people's abdomens are not something that are *routinely* opened; and that human organs are not routinely on display except in anatomical museums; and that the vast majority of us have not seen our own gall bladders. So although my gall bladder is an object, it is essentially hidden, especially to me. Although it is intimately close to me, it is as unfamiliar and inaccessible as the surface of the moon.

When the doctor says that my gall bladder is infected, I do not really understand him. And yet this is not simply a question of technical knowledge or its lack, as when I fail to understand the car mechanic who diagnoses a broken carburettor. To grasp this difference, we have to explore the different roles played by bodies and by cars within our lives. One way to do this is by asking how we teach *children* to use these concepts correctly, for that will reveal certain aspects of meaning that philosophers are likely to miss. What do we respond when Junior asks 'what's a kidney, Daddy?' We could take him to the butcher's, and point: *this* is a kidney, while *that* is a liver (compare the shape, colour, texture, and later compare the taste in the stew). Families who live and work on a farm will be able to show not only the organs of the freshly killed animal, but also their relative location. Either way, the more advanced child can then learn something of the organs' functions, although this will require a more systematic introduction to vertebrate anatomy with simplified concepts and metaphors borrowed from machinery and plumbing. So first the child needs to understand about blood circulation in order for the concept of a kidney as a filter to make sense. Sometime around this point, the child will then be invited to see the connection between the animals and his own body in terms of common anatomy and physiology: the waste liquid from the kidneys of both bodies is called urine; the blood that carries oxygen around the animal's body is the same sort of thing that emerges from the child's scraped knee; the bones left over on the dinner plate are the same things that you can feel under the skin.

However, together with location and function, the child also learns that bodily organs are essentially internal. When the animal is alive, its internal components can never be seen, heard (with the important exception of the heart) or felt (with the exception of the bones and muscles). And here is the key point: insofar as they cannot be perceived, they are not part of the concept of a *living* animal. When asked to draw a cow grazing in the field, the child will not draw its organs through a transparent abdomen as well. In trying to imagine the child's discovery of the world, we have to step back from what we already know and take for granted. A child could be forgiven, for example, for thinking that the living animal might not have organs at all, and just grows them after death, just before the inanimate corpse is opened up. As a result, there is a real sense in which the child just has to take it on trust that he has internal organs, without engaging with the knowledge. The main exception will of course be the heart, which the child can hear and feel, and which thus assumes a primitive place in the child's self-awareness, a place that makes it easier to accept the metaphorical roles allocated to the heart, i.e. the visceral seat of courage, excitement, fear. But even though we have experienced palpitations, we do not know much more about the heart as an organ. Another exception will be the stomach, which can feel full or cramped or suffer indigestion, and indeed which can expel its contents dramatically back into the world. Again, this activity makes it easier to accept the metaphor of having the 'stomach' for a fight. But this proprioceptive knowledge obviously falls far short of the doctor's biomedical knowledge – and this discrepancy will be relevant to our discussion.

What the child's learning leads to, of course, is a grasp of folk medicine, to greater or lesser degrees of accuracy and sophistication. Folk medicine is the sort of thing that newspaper editors and GPs can assume to be known by almost all competent adults. Most medically untrained people can understand well enough what the problem is and what the solutions are when a public figure is said to be undergoing an operation to remove a lung tumour or bypass a clogged coronary artery. And in these days of easy access to information, any of the technical terms can be looked up by inquisitive patients with a minimum of persistence. In addition, folk medicine involves a loose admixture between illness and disease: the former involves the patient's experience (based mainly on feelings and visible signs), while the latter involves the objective description (based mainly on physiological function and defect). Folk medicine will use basic disease concepts to describe the underlying causes of the illness experience.

There is a good deal of discussion in medicine about the problem of communication between folk and scientific medicine, between illness and disease. Medical students now have extensive training in communication skills to help bridge the gap, even if the patient will always have to trust the doctor to a certain degree. But this problem is part of a general communication problem between lay and expert in any field, as we discussed earlier; the layperson by definition does not have the same amount of knowledge, training, experience, skill or judgement to understand the expert fully. Normally the layperson will accept – and ought to accept – an expert statement because he has no rational grounds to doubt it (within certain limits). There will of course be disagreements among experts. But experts are only able to meaningfully disagree if they agree sufficiently on a core of knowledge to constitute their shared discipline. And to such a core the layperson must defer.

Such a picture would make medicine no different from car mechanics, and this is the picture that I now want to challenge. Here I am (with no more than folk medical knowledge), presenting with moderate and persisting abdominal pain. It is not a familiar pain, like indigestion, which I can normally relieve with an appropriate reflux suppressant (and where I can understand the pain and the treatment in a way close to the biomedical model). But on this occasion, the pain is unfamiliar, and there are no superficial abdominal marks that might suggest an exogenous cause. I think it *might* have something to do with a liver or a gall bladder, but here I'm using the words as no more than a synonym for 'abdominal contents'; within my perspective on my own body, there is no essential link between the pain and a particular organ.

After examination, the doctor diagnoses an infection of the gall bladder, and prescribes a course of antibiotics. The doctor, with his communication skills training, explains what has happened: 'There is an organ called the gall bladder, and it stores a chemical called bile that helps break down your food. The gall bladder is under attack from little bugs, and so there is some damage affecting the adjacent nerves, and that causes the pain. These pills will kill the bugs.' (One can imagine further descriptions using more sophisticated biochemical or physiological terminology, tailored to the patient's understanding.) Again, this is the sort of causal explanation a car mechanic might give me, with the crucial exception that it purports to be about *my body*, 'about what is going on *here*, under your ribcage, sir'. The elements of the explanation are not out there, on display, inviting attempts to make sense of them; instead, they are essentially concealed, and concealed deep inside me. Compare this to the problematic carburettor – even if I haven't a clue what it's for – over there, that round, grey thingee on the right, when you open up the bonnet. Once back from the garage, I can show my friend the round, grey thingee.

I want to ask: what is this explanation *doing* here? I'm not asking whether it's true; let's say it is, that the diagnosis is correct and the prescription appropriate. There is nevertheless a sense that most of such an explanation will be, within limits, *arbitrary*. The doctor could say it's the gall bladder or the kidney or the liver: it doesn't really matter to me, as long as the pain eventually goes away (I'm assuming it's a relatively minor, treatable condition). Of course the cause of the pain is not arbitrary at all from the scientific point of view; but the patient does not conceive of his own body in the context of such a point of view. I stress that the explanation is not *entirely* arbitrary: it still has to cohere roughly with the patient's folk knowledge. But within the limits, and while there is no clue on the surface anatomy, one explanation will be as good as any other to the patient. As such, the main point of the explanation becomes rhetorical rather than informative: to reassure the patient by appearing to confidently recognise a minor problem and to declare that a routine treatment is available ... and here it is.

What we are talking about, once again, are two types of knowledge, two sets of meanings, that are applicable in two different contexts. This is not sufficiently recognised as a problem of communication in the medical world, nor is it sufficiently recognised in some of the debates in medical ethics surrounding the donation or sale of bodily organs, for example. An organic chemist could tell me a lot about the composition of a particular piece of plastic, but this would not help me at all to understand what that piece is called in the game of chess, or about how to move it on the chessboard. What the chemist says is not untrue, but it is irrelevant to the meaning of chess; and in no sense can the chemist declare that he has a better, or deeper, or more fundamental understanding of what is going on.

When you tell me that there are 14 maria (seas) on the surface of the moon and when you tell me that there are four chambers in the human heart, I can (within limits, again) accept both statements without demur, and I can accept them in exactly the same spirit, since neither of them are facts about *me*. Indeed, they aren't facts about any people I know. The people I know have heads, shoulders, knees and toes, not to mention closed abdomens, and I regularly come into contact with these parts of people's anatomy, but I've never seen a human heart in a living person, let alone its four chambers. You can tell me the heart has four chambers or 40 chambers, and I can shrug my shoulders and say 'fine', knowing that this will make *no difference* to the way I behave and inter-act with other people outside the hospital, during the vast majority of my life. And, to follow on from the discussion in the last chapter, when I am diagnosed with the early stages of dementia, my fear is based *not* on the explanation of what is happening to the neurones of my cerebral cortex, but on the description of what has happened to the personali-ties and behaviour of others like Iris Murdoch.

And even when I do see a human gall bladder, the diseased one that has just been removed from my friend and which he keeps pickled in a jar on the mantelpiece for his dinner guests to look at, *that* gall bladder doesn't enter into my relationship with *this* person. I don't recognise it *as* my friend's gall bladder, for it bears no distinctive marks of author-ship or ownership. If you swapped it with a cow gall bladder, I wouldn't know the difference. My friend, as I see him and talk to him now, is not different in any way after the removal of his gall bladder – although he might be more anxious about the future than before. But his anxiety has to do with the doctors' warnings and the pills he has to take. The generalised anatomical information about human gall bladders cannot find purchase in the patient's understanding of himself and of others in the world. Compare the place of an infected gall bladder to that of a broken nose. There is nothing particularly special about the latter in anatomical and physiological terms: some component of the machine is again defective. But what a difference it makes *to me*! I can see it, feel it (both proprioceptively and digitally), and other people will see it and ask me about it; when I do visit the doctor, it is clear to both of us what exactly is broken and what he is supposed to do. We might summarise it as follows: there is nothing *uncanny* about a broken nose.

Roger Teichmann (2001) considers the following thought experiment. A friend complains about persistent, but mild headaches. After consult-ing with a doctor, she is admitted for a brain scan. The confirmed results of the scan are quite startling: she has no brain, her skull is filled with fluid. Yet she continues to live and to function quite normally: she responds to pain and other sensory stimuli in the normal way, she has memories that are both internally coherent and cohere with my memories, she can articulate plausible hopes and desires etc. Teichmann's point is that, while scientific medicine might be very interested in the results of this test, it would not affect my friend's understanding of the normal world or her understanding of herself in that world or my understanding of her in the world. Perhaps more importantly, it would not affect my *attitude* towards her: I would not start thinking that she was not *really* in pain when she stepped on a drawing pin and winced.

As such there is a crucial sense in which it doesn't matter what she has in her head: what matters is that she continues to talk and behave as a human being, in ways appropriate to the situation, and manages to communicate meaningfully with other human beings.

This is not to be confused with behaviourism, that is, with the view that there are no mental states but only stimulus-response mechanisms like those of a fly when it wriggles. Teichmann is not denying that the friend feels real pain; what he claims is that the *meaning* of her pain – in terms of interpersonal communication about her pain – has nothing to do with the physiological account offered by scientific medicine, which would require nerves and synapses and acetylcholine and a brain. Perhaps my friend could be an object for further scientific study, and a whole new pain-relay mechanism might be discovered in another part of her anatomy. But whether or not a satisfactory scientific explanation would emerge would again be irrelevant to the meaning of my friend's pain. The meaning of pain, within Teichmann's and his friend's world, has to do with sensations, pain behaviour, situations where that particular kind and degree of pain behaviour would be appropriate, and with the spontaneous reactions of others to a person's pain.[2]

So the familiarity of the pain behaviour overrides the biomedical account in our ordinary lives. Consider the expression 'I knew it like the back of my hand'. This does not mean that I have knowledge of human hands *in general*, although I do. Rather, I have a specific knowledge of *my* hands, and would recognise a picture of them from among hundreds of others. Indeed, they are probably the most visible (to the owner) part of the body during ordinary activities. Importantly, my hands don't just have a certain shape and appearance; they also move in a certain way, they are the medium for particular skills and for particular pleasures and gestures. All this contributes to making the hands familiar, and familiarly mine, in a way that is impossible for bodily organs.

This discussion of the familiar signifies a move away from the problem of expert–lay communication. For I would suggest that even the doctor, even the gall bladder specialist, does not find *his own gall bladder* familiar. Unlike the layperson, the doctor has a large and relatively coherent body of knowledge about human organs; but there is a radical *gap* between this knowledge and his knowledge of people as Jimbo or Karenin, as friend or foe, as tinker or tailor. At some point during the consultation, the doctor switches, more or less abruptly, from the human world to the medical world, from a 'reactive' attitude (to the patient) to an 'objective' attitude (to the defective machine), to borrow the terms made famous by Peter Strawson (1974). The reactive attitude is a spontaneous engagement with the other *as* a human being, a human being with a face and hands and abdomen, but with no gall bladder. Once the torso is opened up on the operating table, then the person effectively disappears and all we are left with is the machine, which is approached with the same attitude as would be taken to a car or to the surface of the moon. As such this is not an epistemological dichotomy: one set of beliefs does not contradict or falsify the other. Instead, both sets of beliefs are held contemporaneously, but only one can ever be foregrounded.

What about the friendly GP, chatting with me reactively, all the while probing my abdomen objectively? Here I would have to say that the GP divides the two objects of attention from each other: he talks to my face (or responds to my voice) while feeling my abdomen, without being conscious of the two objects as being part of the same person. Sometimes the reactive attitude might indeed be inappropriate, as when a male GP is examining the vagina of his patient: there the patient would expect a deliberate shift of attitude.

The tension between these two attitudes is revealed in those uncanny moments when the patient is on his way from one type of object to the other. I can think of a number of striking examples: (i) a recent BBC television programme, in which a brain surgery patient was revived once his skull was opened, so that he could report on the effects of the exploratory proddings by the surgeons; (ii) the Steven Spielberg film *A.I.*, in which a 10-year-old boy robot (played evocatively by a human) was opened up for repair of his mechanical innards while he remained conscious and reassured his adoptive mother (a real human) that it didn't hurt; (iii) the modern penchant for super-realism in war films (e.g. Spielberg's *Saving Private Ryan*), featuring intestines spilling out of surprised soldiers' abdomens; and (iv) the medical student's first scalpel cut into the a dead corpse, or first glimpse of an entire severed head. In all these examples the familiar and the uncanny collide in a startling experience.[3]

Even if the above is accepted, it might be of merely psychological interest. However, I believe the implications go much further. When a patient is asked for his consent to an operation on his 'gall bladder', the patient simultaneously does and does not understand what the doctor is talking about; and therefore can and cannot properly *consent* to the operation. And because he cannot consent in this latter sense, he has to trust the doctor: any talk about empowering the patient, about removing old-fashioned paternalism, about making him a joint partner in the treatment, is bluff to a certain degree. His 'understanding' of the proposed treatment is a mere parroting back of what he has read on the internet, like a child correctly spelling absurdly complex words in an American spelling contest without knowing what they mean. And even after the operation on his gall bladder he will be none the wiser as to what happened. 'They patched up my gall bladder', he'll tell the lads down the pub, but all he really experienced was waking up with stitches on a sore abdomen. For all the patient knows, they could have removed his liver and inserted three cricket balls while they had him open. If the pain goes away, the operation was a success. This is to be contrasted to the consent requested of the patient with the broken nose. Unlike livers, bones make sense to us because we can feel them, many of us know what it means for them to break, and the structural function is close enough to that of building girders. Importantly, I can check (most of) the doctor's work myself. In that limited way, it could be said that we share a common 'expertise' of what looks and feels normal.

The above also has ambiguous implications for the donation of organs after death. It is common to accuse non-donors of being super-stitiously and selfishly possessive about their own organs; such people have not yet grasped that they will not need their organs after death, whatever the fate of their soul. But I would argue that the organs are *not theirs to give or to retain* in the first place. The very concepts of 'possession' and 'ownership' are only applied meaningfully to things that I can create or purchase, or give or receive as a gift, or take into my hand or live in. Even owning something abstract like stocks and bonds involves pieces of paper, bank balances, formal letters and the verifiable knowledge that they can be converted into cash at any time, cash that can be used to buy less abstract things. The failure to understand the peculiar status of bodily organs lay behind the paradoxes surrounding the Alder Hey scandal, to which we now turn.

The Alder Hey scandal

The difference between the human and the biomedical account of the human body was strikingly illustrated by the scandal surrounding the retention of organs at the Alder Hey Children's Hospital in Liverpool. Between 1988 and 1996, over 2000 body parts and many whole organs were excised and retained from the bodies of dead children. In addition, 400 foetuses were stored from abortions conducted throughout the north-west of England. The most serious charge was that the organs had been removed without the explicit consent of the parents. What the parents had usually consented to was the removal of 'tissue', which to a pathologist can of course can be anything from a biopsy to a whole organ. Partly as a result of the Alder Hey scandal (together with a similar scandal at the Bristol Royal Infirmary), the Human Tissue Act 1961 was replaced with the Human Tissue Act 2004. The gist of the relevant clauses of the new Act is to require much more explicit consent from the parents or relatives for the retention of any human tissue, be it biopsy slides or blood samples or even faeces, but more importantly explicit consent for each bodily organ.

Clearly there were real problems with the procedure in place at Alder Hey, but it is not clear whether the new Act will resolve all the underlying ethical issues. The most obvious problem is that much less tissue will now be retained: not only because of the cumbersome paperwork, but also because families will be less likely to give the requested explicit consent. This will be a huge detriment to scientific research. Now in response it is tempting to say: 'too bad'. There are other things that would benefit scientific research enormously but that are not permitted anyway: the lethal experiments conducted by Nazi doctors and scientists on concentration camp inmates, for example. Science does not need to advance at all costs. I do not want to get embroiled in this particular debate, but clearly there has to be a trade-off.

What is of greater interest to me, and to the broader issues raised so far in this chapter, is how the two different discourses of people and bodies have clashed on this occasion, where I use the word 'discourse' to comprise interrelated ways of thinking, speaking and acting. On the one hand we have the whole cultural discourse surrounding dead bodies: the elaborate rituals of mourning and wakes, the dress codes and musical genres, the respect due even to the bodies of dead strangers, the flags at half mast and gun carriage processions for dead public figures. On the other hand we have the powerful biomedical discourse involving the body as no more than mass of cells and chemicals. After all, the pathologist will tell us that there is nothing essentially different between a live and a dead body; for when a body dies, it is simply a matter of certain physiological processes coming to an end and of others beginning: there is no person left 'in' the body, and the *entire* cadaver is 'tissue': there is no essential difference between some flakes of dead skin, the non-functioning gall bladder or heart, or the dead leg or face.

The clash of discourses can be seen in one response to the 2001 report into the Alder Hey scandal.[4] Dewar and Boddington (2004 p. 463) criticise the wording of the text for playing too readily into a tradition of 'the macabre, horror genre of film and literature, and the real history of grave robbing and body snatching'. This encourages a 'fantasy of residual feeling or human sentience in the dead, or parts of the dead' (Ibid. p. 464). Indeed, the *Report* (p. 19) sympathetically cites some of the Alder Hey parents who speak of their children being 'stolen', or who describe the whole episode as 'feeling like bodysnatching'. The biomedical discourse would entail a criticism of the Alder Hey parents not just for selfishly hindering scientific progress (and preventing the ensuing benefits of the research to other children), but also for being irrational and superstitious for denying the truth and clinging to their child's body and the body parts, after the child has gone. After all, what would the parents *do* with the heart that they receive back? Most likely allow it to decompose after burial, or have it cremated: they certainly cannot display it like a family heirloom, or handle it in most ways implied by the concept of ownership.

Mary Warnock describes the Alder Hey parents thus:

> The violence of the reactions was surprising. What did those furious parents believe about their dead children? Did they think that in some way the future of the child had been compromised by the removal of the organs? If I had been such a parent I would have been angry at not being consulted. I can imaging raging against the arrogance of those doctors who seemed to believe that the child's body was their own property, and who thought of the dead child as so much research material. But though such anger is intelligible, is it rational?
>
> (Warnock, M. 'The death of reason' in:
> *The Guardian*, 2 September 2006 p. 25)

This paragraph is very revealing and worth examining closely. First of all, the usual emphasis on rationality as the guiding criterion for evaluating other people's actions. It is hard to imagine Warnock *comforting* the grieving parents, all the while thinking to herself 'this behaviour is irrational'. Indeed, she might see herself as doing them some good by admonishing them in a schoolmarmish sort of way: 'For God's sake, be rational, it's just a dead body.' Indeed, even suggesting to the grieving parents that the organs are mere tissue and that they could be put to useful purposes seems to compound the insult. As I discussed in Chapter 7, the criterion of rationality does not seem to fit very as well into events surrounding death as it does into events surrounding the buying and selling of houses, where it is directly related to a clear conception of self-interest and the future. But it is hard to see what sort of self-interest would be at stake in one's spontaneous grief at the death of one's child. Instead, if the parents are to allow the removal of their child's organs, they have to conceive of them as something like a gift; and a gift by definition cannot be expected or demanded on the grounds that refraining would be irrational.

What is interesting is that Warnock tries to downplay this cold-heartedness by conceding that she too would be angry if she had been in the same situation. But although *her own* hypothetical anger would be intelligible, she implies, it would not be rational. This is a rather odd implication. We can omit the above problems of invoking rationality here by speaking instead of justifiability, but why think the anger would not be justified? If she genuinely believes that the scientists' behaviour was deceitful and arrogant, and it seems fairly clear that it was, then surely there is sufficient justification to be angry?[5] Now there are some situations where one is intelligibly angry, but on reflection one later realises that one overreacted: I resent being given a speeding ticket at the time, but later I reflect that a system of speeding fines is probably a good thing, and I was in fact speeding. But this situation is not the same as the Alder Hey, because the essential presupposition of the 'system' that Warnock is asking us to rationally support – that bodies are no more than useful tissue – may simply be unacceptable when it comes to one's *own* child, not only at the time of death, but ever.

And yet Warnock accepts that 'it is right to treat the death of any human being as an occasion for the formal expression of grief and respect, the public recognition of loss'. But what does this acceptance mean, exactly? It is a description of the psychological response of the parents, a response that is intelligible insofar as being shared by most of us. But Warnock and the parents have different conceptions of the *object* of that response. For Warnock, they are grieving, intelligibly and appropriately, for the *past* child whom they will never see again. The rational thing to do is then to gradually accept that the child is no more, that the body of the child is now empty and may as well be used to benefit others. For the parents, however, *the child still exists, even if he is dead* – they are grieving not for a past child, not for an image, not for a memory, but for their child, this child *here*.

This sounds a bit spooky, if not downright odd, so let's pause. It is not controversial that there are many different kinds of existence. Tables and chairs, tissue and cells are the most obvious kind. The existence of beauty, the past, numbers, quarks and human minds is a bit more abstract to describe, but still undeniable in ordinary lives. But what could it mean to say that the child exists even though it is dead? I'm certainly not talking about a ghost or about a heavenly resurrection here. Instead, I am returning to one of the themes of this book, which has been to argue that individuals are not isolated atoms with contingent links to other atoms; instead, if two human beings have a close relationship then that relationship will survive the death of one of its members, and the dead will continue to exist in the living.

Imagine that one year after the child's death, the parents visit his grave and lay flowers on it. While they are there, they start talking about how his siblings and cousins are getting on, telling him that they miss him, love him and hope that he is doing well. Again, there need be no mention of an afterworld at all. What are we to make of such behaviour? On the one hand, the parents *know* that the child is dead, and that his body has greatly decomposed under the grave. Does this mean that their behaviour is irrational and superstitious? It might be, but that wouldn't be a very interesting conclusion. Again, my first impulse is to take seriously what ordinary people do; if the philosophers or the scientists cannot make sense of what ordinary people do, then so much the worse for the philosophy and the science. Why not say that the parents are talking to their child, and in *that* sense – in the sense that he is the person they are visiting, the person they are talking to – he exists? If you ask the parents whom they are talking to, they will *not* say 'to the grave of our son', or 'to a projected fantasy of him', or 'to the memory of him'. Graves, images and memories are not the sort of thing that one talks to, people are.

Compare this example to a case of genuine superstition. If an employee refused to come into work on Friday the 13th, the employer would be rightly annoyed. And yet the employer would be much more sympathetic if the employee claimed that he had to visit his child's grave on the anniversary of its death, and so could not come in to work. It would be crassly inappropriate for the employer to say 'can't it wait until the weekend? After all, he's not going anywhere'. The difference lies in the reality of the relationship between the parents and their child, whereas no such relationship exists with Friday the 13th.

The post mortem

In the previous chapter 1 introduced the concept of 'posthumous harm', with the classic example being that of Achilles slaying Hector and then dragging his body around behind his chariot. Hector's body is undeniably damaged by such behaviour, but is *Hector* harmed by it? That is, harmed not so much by the stones and pebbles, but by the disrespectful actions? (For those who are confused by talk of harm in this context, we can ask whether the disrespectful actions somehow *reach* or *touch* Hector.) Most utilitarians and most writers in medical ethics tend to equate harm with the physical or psychological suffering of the living, and so would throw out the idea of posthumous harm, and with it any of the suggestions I have made about posthumous existence. They would of course allow for family and friends to be distressed by the sight of the treatment of the body, but such distress would not relate to the dead person. What is important is that if the on-lookers really believed that Hector's body was an empty shell, they would be no more distressed by its mutilation than by a painting of Hector being slashed.

Perhaps the Alder Hey parents did not object so much to the removal of organs as to the necessary mutilation of the body required to get at the organs. This might also be a reason why people do not sign organ donor cards, even when they do not really care much about their organs. The question then becomes whether it is acceptable to fear posthumous mutilation in this way.

In an article arguing for a 'rethink' in our attitude to the bodies of the dead, Julian Savulescu opens with a moving description of his 87-year-old father's death. As a doctor himself, Savulescu was not satisfied with the explanation of the death, which involved a jaundice of unknown origin and a suggestion of mismanaged bleeding. So he requested a post mortem. Both in medical school and in professional practice, Savulescu has witnessed a great number of post mortems:

> I learnt an immense amount from these activities. But I also knew how gruesome the autopsy is. I knew that an autopsy would mean that my father would be dismembered. But I had no hesitation in requesting an autopsy. Both I and my mother accepted that his body was dead. He would not be harmed. And important knowledge would be obtained.
>
> (Savulescu 2003 p. 127)

The rest of the article follows a familiar path: bodies are not people, and therefore we should accept the removal of organs that can provide plenty of benefit as transplants or research specimens. What is distinctive about Savulescu's article is that the body in question is that of a person whom he knew and loved dearly, and this is what I want to explore in this section.

First of all, we should be clear just how gruesome a post mortem is. It is not simply a discrete cut in the abdomen, followed by a quick peek inside with a torch. Instead, the abdomen is opened wide and the entire contents removed and placed on a bench, where the organs can be examined one by one by the pathologist. The top of the skull is sawn off, and the brain too is removed and examined. All the organs are then put in a plastic bag and sown back into the abdomen. Gruesome is indeed the word – and yet all of this is legal, and there are good, sensible reasons for having this practice in place. But somehow the good sensible reasons don't do much to alleviate the gruesomeness. Indeed, it is much more than gruesome; it involves possibly the most spectacular defilement of the human body possible. I choose the word 'defilement' carefully; in any other context, the actions of the technician would constitute the most revolting of assaults, worse than actually killing the person: the abdominal cavity wide open and empty, the naked limbs and genitalia splayed out beneath, the skull gaping open and empty.

Now Savulescu knows better than I do what a post mortem looks like. And yet he has 'no hesitation' in requesting one. On the one hand, his father was dead and *he* could no longer be harmed; and yet in an interesting slip, he says that his father (not his father's *body*) would be dismembered. Does Savulescu not acknowledge the merest possibility of defilement? After all, if we were to hold Savulescu at his word that no harm was involved, that it was merely a dead body like the many others he had examined and dissected, it would not be at all strange to suggest that Savulescu be present at the post mortem or, indeed, that he conduct the post mortem himself, provided his technical skills are up to it. For who better to ensure that the job was done properly as well as respectfully than a loving son?

But such a suggestion would be appalling. While we would acknowledge the importance, even the right, of Savulescu to see his father's body for 'one last goodbye', the body would of course be carefully prepared for viewing, to give the illusion that he was merely sleeping. But Savulescu is a medical man, presumably lacking the squeamishness of the rest of us, so what if he did claim the right to observe the post mortem, even despite advice that it would be disturbing?

What is also revealing about Savulescu's article is that he does not mention his *father's* wishes. His father might well have signed the organ donor card, but would his father want to have been dismembered?

Indeed, would he have wanted to be dismembered by his own son? None of these questions should be appalling, or even embarrassing, if we take Savulescu at his word that his father is no longer in the body and cannot therefore be harmed or defiled or disrespected. My point, of course, is not at all to argue against the necessity of post mortems, but to suggest that there is a real paradox that Savulescu cannot so blithely 'define away' by declaring the supremacy of the biomedical discourse of the body and ignoring the cultural.[6]

10
The Euthanasia Debates

If one is to judge by the sheer volume of literature on the subject, euthanasia is perhaps the biggest topic in medical ethics. Indeed, it is too big for a single chapter, and so I have already discussed some of the relevant issues in previous chapters. However, I have divided the topic up not only for such arbitrary logistical reasons, but also in accordance with the methodological principle of this book. I believe that there is a lot of groundwork that needs to be covered before one can even begin to talk about euthanasia seriously. Just as one needs to understand something of the place of normal human birth in human lives before talking about abortion, so too does one need to understand something of the place of normal human death in human lives, before one can discuss suicide and euthanasia. Even with such ground-laying, the topic remains vast, and I have to satisfy myself with no more than a few gestures in this final chapter, picking up some earlier themes.

My position, insofar as I have one, is nothing so crude as to 'support' or 'oppose' euthanasia – even assuming we can all agree what we mean by euthanasia (I return to the definitional point in a moment). Instead, I claim two things. First, that some of those who debate the merits of one or the other position in the classroom or in the media often get so lost in the theory that they have not truly understood what it *means* to die, that is, (i) what it means for an individual to 'properly' contemplate his death or the death of those close to him and (ii) what role is played by death in our individual lives as well as in our society as a whole. Second, that it is very hard to know what one's position on euthanasia (insofar as anyone can or even ought to have a coherent single position) is until one has been in the situation where one has to make up one's mind for real.

And just as I argued in Chapter 5 that men did not have the same authority to debate abortion as women, so too do I claim that, *prima facie*, the young and middle aged do not have quite the same authority to debate euthanasia as the old. I say *prima facie* because of course some young and middle-aged people have earned the appropriate authority on the basis of bitter experience. The death of a loved one, or one's own near-death, is much more important for true understanding than armchair philosophical reflection on death. But even the experience of another's death might be insufficient to prepare one to actually make a generalisable decision about whether to help another person to live or die, when there are so many contingencies at stake. It is one thing for a terminally ill wife to ask her husband to promise her that he will help her to end it all quickly once the pain gets too much; it is quite another for the wife to ask the husband to put a pillow over her head *here* and *now*. There need be no shameful inconsistency or weakness or selfishness in the husband sincerely promising to help and then finding that he is unable to go through with it.

Similarly, young people contemplate the decay of their elders with horror, and swear to themselves 'when I get like that I'll just jump off the bridge'. And yet what is interesting is that relatively few elderly people commit suicide,[1] even when they reach a stage or have been put in a place which they had earlier vehemently believed to be intolerable. The fact is that people can learn to live with the worst conditions. Whether such toleration is to be condemned or admired is a secondary question. The statistical probability is that most of us will not *seriously* contemplate suicide or *seriously* request euthanasia once we become elderly and infirm. In this very specific sense the public debates may be somewhat exaggerated. Nevertheless, there is a real ethical question about how to respond to the serious few. And a doctor's or a society's attitude to those few can reveal a good deal about their general ethical outlook, and this also makes it worth discussing.

First, in line with the critical aspect of this book, another cautionary tale. A leading textbook on medical ethics gives the following case scenario in their chapter on euthanasia:

> A driver is trapped in a blazing lorry. There is no way in which he can be saved. He will soon burn to death. A friend of the driver is standing by the lorry. This friend has a gun and is a good shot. The driver asks this friend to shoot him dead. It will be less painful for him to be shot than to burn to death. Should the friend shoot the driver dead?
>
> (Hope, Savulescu and Hendrick 2003 p. 162)

Given my response to McMahon's scenarios in Chapter 1, it should be obvious what I am going to say in response to this one. The example is so contrived and so vague as to be utterly pointless, and any seminar discussion provoked from such an example will be just as pointless. But my criticism is not just a point about detail. Even if the scenario were presented as a longer story, with all the details carefully worked out, there is a massive difference between asking the comfortable audience whether the *friend* should shoot the driver, and trying to imagine what *I* would do in that situation. The case scenario asks for advice, and such advice comes too cheap from people who are not implicated in the horror of the situation, who will not have to live with the decision they made in that situation, and who probably don't know much about death and dying and killing anyway (assuming the book's main readership to be callow students). And whatever advice I would like to give the 'friend', there is a sense in which I haven't the foggiest idea what I would do *until* I got into that situation, and maybe all I could say to the friend is 'God help you'. Even if more detail could be presented by means of a skilfully acted film, this would still be woefully insufficient to capture what it would be like for me to be in that situation, and what it would be like for me to have to make such an awful decision within the context of my relationship with the driver. And then the situation itself will be so hectic, so urgent and so horrible that it would be utterly implausible to speak of a *reasoned* or even *reasonable* decision – instead, the decision would make itself in the heat of the moment; I would do what I felt I had to do, and my whole life would be transformed in radically unpredictable ways.

The word 'euthanasia' comes from the Greek for 'good death', but this does not help us very much, since there are so many types of patients and so many ways to die. One component has to be that the death is *hastened* in some way, either in comparison with the expected lifespan under the untreated progression of a disease, or – just the opposite – in comparison with the expected lifespan under available treatment. A second component, the 'good' part of the hastened death, has to involve a belief that such a death is a benefit to the patient, where this is usually to be interpreted as a release from present or anticipated suffering and indignity. So far the notion of a good death makes no distinction between acts and omissions that knowingly hasten death (the distinction sometimes referred to as that between 'active' and 'passive' euthanasia); makes no distinction between the doctor and the patient as final causal agent of the death-hastening; and makes no distinction between competent and incompetent patients (a distinction sometimes referred to as that between 'voluntary' and 'non-voluntary' euthanasia). Needless to say, if the patient is competent, then euthanasia requires the patient's free and informed request.

Most of the debates surrounding euthanasia focus on the competent patient's request for active euthanasia, as in the case of Diane Pretty in Chapter 7 (although technically she was requesting permission for her husband to assist her suicide). Yet my wider definition would include cases such as Ms B and Tony Bland,[2] which are not ordinarily defined as euthanasia (indeed the Bland ruling explicitly denied that this was a case of 'euthanasia'.) But it will be useful to recall the details of these two cases.

Ms B was a woman in her 40s, irreversibly paralysed from the neck down and relying on ventilator support (among other things) to stay alive. She competently requested that the ventilator be switched off, she was warned that she would no longer be able to breathe and that she would die, she accepted the warning and repeated the request. The hospital trust took the matter to court, on the basis that she was incompetent because of temporary depression. The court overruled the trust and declared that removing the ventilator would be lawful. (Interestingly, the court did not *instruct* the Trust to switch off the ventilator, even though not doing so would be continuing to assault her; the Trust refused the court's ruling, but arranged for Ms B to be transferred to the care of another Trust who were willing to switch it off – which they did, and she died.) Therefore, Ms B competently requested an omission which she knew would kill her. The key question here was whether she was depressed, and therefore whether her competent judgement was impaired. Once it was established by the court that she was not depressed and therefore sufficiently competent, her clear and consistent wishes for the removal of treatment – *any* medical treatment – had to be respected, in accordance with a principle that is very strong in medical practice.

The definitional question then returns: was this a case of suicide? And if so, were the doctors assisting that suicide, and therefore making it a case of euthanasia? And yet the case is not normally described as one of suicide or euthanasia. Compare this to Ken Harrison, which seemed a much clearer case of both. And yet the only difference between the two cases is that Ms B required a ventilator to stay alive: this is a clear example of a medical intervention, an extraordinary device that one would not encounter in any other circumstance. All Harrison needed was food and water. Harrison could have refused these by simply closing his mouth to the proffered spoon. Eventually, however, Harrison would have lost consciousness, and the hospital would then feel obliged to feed and hydrate him intravenously or by a PEG tube, in effect force-feeding him. The point here is that artificial nutrition and hydration (ANH) is normally not considered medical treatment but 'basic care'. Basic care comprises a set of procedures designed to keep the patient comfortable and clean, and these can never be discontinued, even at a competent patient's request. This is because discontinuance would result in distress to the visiting relatives, staff and other patients, but perhaps more importantly, the thought is that something like hygiene is important to human dignity even when the patient himself has lost all interest in hygiene or, indeed, in dignity. In addition, eating and drinking are often not done in private, but have an important social value, and so making sure the patient is properly fed and watered constitutes an indirect effort to keep them in the social world.

As one can imagine, some utilitarians would find the distinction contrived, just as Peter Singer did not distinguish between Diane Pretty's request for a poison and Ms B's request for the withdrawal of the ventilator (see Chapter 7). In all cases the results can be reasonably foreseen, and that equal foreseeability makes the parties equally responsible for the result. However, in terms of my interest in the wider meanings of ethical terms, I do accept the difference between active and passive, and between withdrawing medical treatment and withdrawing ANH. The latter distinction, however, was notoriously challenged by the Tony Bland case.

Tony Bland was injured at the Hillsborough stadium disaster in 1989, and ended up with severe brain damage and in a coma or persistent vegetative state (PVS). However, Bland differed from Ms B in *not* requiring a ventilator or any other medical treatment, and differed from both Ms B and Ken Harrison in being unconscious and therefore unable to express his wishes (and he was not known to have made an advance directive). Four years after the accident, the PVS still seemed set to continue indefinitely, and so the medical team, with Bland's parents' support, asked the court for permission to withdraw the artificial nutrition and hydration in the full knowledge that he would die as a result. Permission was granted, ANH was withdrawn, and Bland died. However, the courts in *Bland* argued that because Bland could not swallow, the nutrition and hydration had to be administered via a naso-gastric tube; this tube could only be inserted by trained medical staff, and pouring the liquid down the tube could not be considered *eating* and *drinking* in the full social and cultural sense, especially since the unconscious Bland was not choosing to eat or drink, and the nutritional mixture that was being poured in did not resemble any known food in taste or smell. Therefore ANH could be classed as medical treatment, and so it could be withdrawn under the two general conditions for withdrawing treatment from the incompetent: that it is futile, or that the anticipated benefits of the treatment are outweighed by the anticipated burdens of administering it. Let me take a small detour from *Bland* to discuss these notions.

Futility, best interests and arbitrariness

The great advantage with 'converting' the matter into a question of futility (or a benefits/burdens balance) is that it is thus effectively transformed from a public ethical matter into an almost entirely medical matter. Unless the patient or the relatives are themselves medically trained, their authority to challenge medical opinion is greatly reduced. But in the case of Bland things are more complicated than they seem.

First of all, there are actually two kinds of futility. As far as I know, orange juice has no known curative effect when used as a treatment for malignant cancer. Orange juice thus constitutes an utterly futile treatment, and the clinician can be under no legal or ethical obligation to administer it, no matter how vehemently the patient demands it. Since orange juice is relatively cheap, however, there might be psychosomatic reasons to 'administer' it to the patient after the failure of mainline treatments, if the patient demanded it. (There is a separate question about whether to prescribe more expensive, but scientifically more suspicious treatments such as homeopathy, but I do not want to get into that.) More relevantly, the second kind of futility is where there *is* a certain small probability of a success, where the probability can be calculated inductively on the basis of past treatment in similar cases. But here we require a threshold below which the treatment would be deemed futile and therefore would not have to be offered. We also require a criterion of success, since the treatment may succeed only by removing the immediate crisis but without allowing for discharge from the hospital.

And this is the important point: both the threshold and the criterion of success can be *legitimately* challenged by a patient who is desperate enough. Even a threshold of less than 1 per cent, and a success involving no more than the postponement of imminent death may be 'good enough' for the patient to demand it. By 'legitimate', here I mean that the challenge makes sense within the patient's perspective of the world and of the bleakness of his prognosis; there is no reason why the patient should accept the doctor's explanation of why he is to be denied the drug, so long as it is not completely futile. Many routine decisions by doctors cannot be sufficiently understood by the patient, and so the patient simply has to trust the doctor. Any protest by the patient is thereby illegitimate, and the doctor can end the conversation by saying, 'if you want to get better, then take these pills – you have to trust me that they will work'. However, when it comes to the alleged futility of a potentially life-saving treatment, once the patient knows the probability statistics, then he is capable of making a desperate but legitimate protest, which may resist attempts to persuade. The determined patient can then only be refused by 'force', by which I mean rhetorical stonewalling ('I'm sorry, I'm just not going to give you the drug because I think it's too futile') or hiding behind an arbitrary policy ('I'm sorry, the Trust has decided on 1 per cent as the cut-off, and I can't go against that').

In Chapter 7, I argued that the concept of rationality only made sense within the context of an on-going life, and could thus not be ascribed to or withheld from a decision to commit suicide. The same point holds true with the concept of futility; it becomes idle when used to justify the denial of something that could save life itself. Futility is normally about the use of an instrument to achieve some goal. If I want to build a fence, a hammer will probably have instrumental value, whereas a feather will be futile. If I want to get a job at Oxford teaching economics, and I do not have a doctorate in economics, then applying for the job would be futile. Of course I have *some* chance of getting the job solely in virtue of submitting an application, for maybe nobody else saw the advertisement and maybe they are looking for other redeeming virtues which I possess. The assessment of the probability is in a sense up to me; it is my time and energy that I risk wasting. But both the decision about the fence and the university job involve an implicit reference to the future, where I imagine myself being alive to enjoy the fence and the job, or to regret having wasted time on both. Whereas if it is my very future that is at stake, it makes sense to demand *any* treatment that has even the smallest chance of success, and thereby to legitimately challenge the doctor's evaluation of futility – for what have I got to lose?

A similar point holds with the benefits–burdens balance, although here the arbitrariness is slightly more evident. After all, it is the *patient* who will experience the benefits and the burdens, and this suggests that he should indeed have the final say about whether the proposed treatment is worth trying. This often happens in practice, where the doctor may be genuinely ambivalent and so will offer the treatment with a sincere attempt at fully informing the patient of the likely side effects. But I am not talking about such decisions, but rather about those situations where the likely benefits are considered so minor and/or so unlikely, and the harms so great and/or so likely, that the doctor considers the treatment futile and *therefore* refuses to even offer it (or refuses to accede to a request to administer them). Again, such a decision by the clinician can only be arbitrary, when seen from the perspective of the dying, desperate patient.

Back to *Bland*. The concept of futility (and benefits/burdens) is invoked to justify a refusal of treatment to a competent patient who is demanding it. It can also be used to justify a withdrawal or withholding of treatment to an incompetent patient, although in practice this usually involves some reference to the patient's best interests. Thus a treatment is futile if there is too low a probability of it serving the patient's best interests. This is another concept that is a lot more slippery than would seem to be indicated by the confidence with which it is used to forestall debate over treatment decisions. For a start, it is not clear what the 'best' adds to the 'interests'. The *best* interests of the patient, i.e. the option that would be best for the patient, would surely be to have the entire NHS budget re-allocated towards efforts to treat and cure him! Anything less than that will be less than the best, and once again we have a problem with arbitrary thresholds.

The decision to withdraw ANH from Bland was explicitly justified on the basis of his best interests. Let us examine this concept. As with rationality and futility, the concept of 'best interests' is essentially future oriented, and only makes sense within the context of an on-going life. It is in my best interest to have my broken leg set in a plaster cast in order for it to mend and for me to be able to use it again in the remainder of my life. On this understanding of the concept, it can never make sense for the person in whose best interests we are claiming to act, to be killed (or deliberately allowed to die), i.e. to be denied the possibility of even having best interests. One can accept that being in a PVS is perhaps not the richest of lives, but *given* that Bland is in a PVS, he does not appear to be suffering enormously, surely it will always be in his best interests to be kept alive rather than to die. Similarly, the ANH that Bland had been receiving could not have been futile, since it was 'working' very well, it was doing everything that it was supposed to do, and that was to keep Bland alive. Indeed, among medical treatments, ANH is probably one of the most successful and therefore least futile, whenever it is used.

And in terms of the balance of benefits over burdens, then after the insertion of the NG tube, it causes no known discomfort at all to a PVS patient, while providing the greatest of benefits, continuing life and continuing hope.

However, maybe I am missing the point in an important way. Maybe with Tony Bland, the point was not that it was the *treatment* that was futile, but that it was Tony Bland's *life* that was being deemed futile – in the sense of being worth less than the lives of the rest of us. This is John Keown's forceful interpretation of the *Bland* decision (2002). And this brings us full circle back to the attempted definitions of personhood, with which I began this book. Because Bland had lost some of the capacities that are essential to personhood – such as consciousness, a sense of the past and future, the capacity to form and maintain rich relationships, take your pick – he was that much less of a person, and therefore 'worth' less than the rest of us full-blooded persons when allocating scarce resources. But even that is not quite right, since ANH is not particularly expensive in NHS budgetary terms. It is not simply that Bland's treatment was insufficiently cost-effective; rather it was argued to be wrong, full stop. He had become a non-person, with no likelihood that he would ever recover personhood. And since he was a non-person the ANH was not even doing its job of 'keeping a person alive'. Instead, it was concluded, there was a shadow of the former Tony Bland, and the ANH was prolonging the limbo between life and death, preventing the family from grieving, preventing a tidy last chapter to Bland's story, preventing the case from being properly closed.

One can accept much of these last few lines while still disagreeing that Bland ought to be allowed to die. It only becomes paradoxical if we accept the utilitarian assumption that the end, described impersonally, justifies the means, which involve a strong personal commitment. John Finnis, in the context of a position similar to Keown's, puts the point well:

> It can be perfectly reasonable to *feel* that death would be a welcome relief for someone suffering from hopeless debility or illness, or from intense and intractable pain, and to *wish* for that relief from suffering which death promises to bring. It cannot be reasonable to form the *judgement* that, all things considered, this person would be better off dead, or the world would be better off if this person were dead, or this person is someone who *should die*.
>
> (1995 p. 65, Italics in the original.)

It is worth stressing that neither Keown nor Finnis are 'vitalists', that is, neither are determined to save lives at all costs and regardless of the patient's wishes. They still allow for life-saving treatment to be refused by competent patients, and they still allow for certain genuinely medical treatments (such as antibiotics) to be withdrawn from dying patients on the grounds of futility. Indeed, it is also acceptable for nutrition and hydration (natural or artificial) to be withdrawn when the patient is clearly dying, and has lost all appetite. However, Bland was not dying, and so there was a chance, albeit very small, that he would eventually wake from his coma – even though he would be severely mentally disabled because of his brain damage. But the most likely outcome would be for him to continue in his comatose state for many years to come.

Implications of Keown's position

It will be obvious by now that I agree with most of Keown's position, but I still have difficulty with some of its central implications. Keown clearly thinks the courts made a mistake in allowing the removal of ANH from Tony Bland, mainly because this implied that Bland's life was of little or no worth, but also because it amounted to giving up hope. And he clearly believes that redefining ANH as medical treatment was a contrivance to rationalise a decision that they had already made about the worth of his life. But let us look at Bland's diagnosis and prognosis for a moment. His cerebral cortex, the site of all higher cognition and personality, had liquefied; that part of the brain would never return to normal function. What was still alive was the brain stem, which controls the primitive organ systems such as the heart, lungs and digestion. There might have been a possibility that he would regain rudimentary consciousness as a severely neurologically impaired individual. In itself, that might have been enough to continue treatment. But again, the chances of this much recovery were extremely slim: he had been in PVS for four years by the time of the trial. And yet in terms of the health of his organs, there was every likelihood that he could be kept alive on ANH for another 30 or 40 years in that PVS.

Is *this* what Keown was advocating? Bland lying there in a bed, with a brown liquid being passed down a transparent tube into his nose; being shaved and wiped and bathed and clothed by nurses and carers; with his parents and some of his former friends visiting every few days, chatting with him, holding his hand, and then leaving – and without *any* reaction from Bland, day in, day out … for decades? At least the demented Margo seemed to be enjoying life and was still able to gesture and speak in ways that reminded others of her former self.

The *Bland* decision crucially depended on the parents' agreement; if the parents had refused to allow the removal of ANH, this is how Bland would have continued to live, and this is how hundreds of other patients live in the UK today. There might eventually be a new decision to be made if Bland were to contract pneumonia. A straightforward medical remedy exists, but it might be easier to withhold the antibiotics precisely because they constitute an unambiguously medical intervention – unlike the ANH which is less plausibly considered medical. However, Keown would be just as adamant in rejecting this course, unless Bland were already dying. How far would Keown go? I suppose until Bland got some sort of more serious disease or condition such as cancer. But I have to say it again: this might take *decades*. Is that really what it means to accord Bland the full respect he is due?

Now consider again the case of Ms B. According to law a competent and informed patient can refuse any medical treatment (including a ventilator, in this case), even when it is life preserving. The court case, remember, turned on whether she was competent. But Keown ('The case of Ms B', 2002) considers another aspect, and that is the underlying motive behind the refusal. If the patient's intention is to commit suicide *by* refusing treatment, then the doctor would be assisting her suicide *by* removing the ventilator. This Keown would not countenance, for much the same reasons as before: it would involve an implicit agreement between the patient and the doctor that the patient's life was not worth living. He would accept the refusal of sufficiently burdensome treatment, however.

But a ventilator is not particularly burdensome, *given* the paralysed state of Ms B's body. Being on a ventilator is certainly burdensome compared to Ms B's previously active lifestyle, but that is the wrong comparison to make. The accident has happened and Ms B is paralysed; now is when we have to make the decision about what to do, and to compare options. Similarly, *given* that she is paralysed, the ventilator offers a massive benefit: continuing life. The rest of her is in good enough shape that her prognosis might again be two or three decades. As such, there is no way that she could justify removal of the ventilator to Keown.

The problem here is a certain arrogance of power. We can imagine Keown in the room with Ms B, having almost absolute power over her. Can we not imagine Ms B's *fury* at Keown for telling her that her life is in fact worth living, whether she agrees or not, and that her suicidal thoughts are merely from depression? Ms B does not have an equal bargaining position as Keown, if you will. She is entirely dependent on those around her for the fulfilment of every last intention. Her sense of helplessness and isolation would only be made worse by this most paternalistic of judgements: to keep someone alive, who carefully, soberly and insistently had decided that she did not wish to live in that condition. It is tempting to speak of the fury as that of the prisoner, with the doctor as the gaoler, but this analogy is far from useful. The genuine prisoner longs for a *life* outside the prison, where he will be himself and resume his projects and relationships. Prison is meant to be a miserable experience precisely because of that disruption. Ms B longs for life outside her ventilator, but not in a way that the doctor can do anything about. As a result, she longs for death. She has evaluated her present life, its likely future, and she has decided that it is time to go, thank you very much. In this way her situation is not much different from that of a very old patient waiting to die in a hospice, but it's taking longer than anybody thought. Here there need be no thought of refusing treatment, merely of thinking to oneself that one is ready. And that every moment *after* that decision assumes the distinctive indignity of a lingering guest. When such lingering is directly controlled by the doctor figure by an act of will, this can lead to a desperate fury: how *dare* he arrogate such power?

I confess to being torn here. My sympathies are with Keown, and certainly my discussion of dementia, in Chapter 8, also suggested a resistance to euthanasia, but I am struck by the awful prospect of Bland going on and on, of Iris Murdoch going on and on, and of Ms B being forced to live, on and on, because she happens to lack the means to kill herself. However, I am also reluctant to choose sides because I lack the appropriate authority, and would prefer to leave it in the hands of the relevant doctor. This is in itself controversial, for it implicates the doctor in a business which goes directly against the doctor's healing mission. Surely if someone is to be deliberately killed or allowed to kill oneself or allowed to die, the matter should be decided by something like a tribunal, in the same way that decisions about capital punishment are made in certain countries like the United States?

However, I would have to support the present 'fudge' of the British system, and with the role of the doctor within it. It is worth spending a final section discussing this fudge.

The symbolic and the regulatory role of the law

At the heart of the present English law is a fudge: the known gap between theory and practice.[3] Officially euthanasia is prohibited in the strongest terms. In practice, however, the law regularly turns a blind eye to the occasional dose of an analgesic that will knowingly hasten death. We need to understand the difference between a fudge and discretion. Discretion is where the law *cannot* by nature be any more precise. A policeman on the beat is given certain powers, e.g. to stop and search, and in his training he will receive *general* guidance on the sort of circumstances that would legally justify a stop and search. But it is impossible to specify every possible circumstance, to specify one set of circumstances down to the tiniest detail, and to expect the policeman to compare the situation before him to that in the rulebook so precisely. So the policeman has to use his discretion, sometimes called his judgement. The quality of his judgement will also depend on the judgement he is forced to exercise in ordinary non-professional situations, but above all he will be helped by the quantity of experience on the job, and by observing experienced colleagues 'get it right' even when they lack sufficient explicit evidence or guidelines.

The reason that the euthanasia fudge is not a question of judgement or discretion is that the law *could* be made much more precise and *could* be enforced much more effectively. But the thought behind the fudge is that it is somehow better if it remains as an unqualified prohibition. Interestingly, the current fudge can be criticised by both opponents and proponents of euthanasia: opponents claim that a vague law allows too much euthanasia; proponents that it does not allow enough of the right kind. Surely *any* law, both sides argue, if there is to be any point to it, has to be made as precise as possible and enforced as much as possible, otherwise it would just be a bit of paper. Transparency and consistency are hallmarks of any legal system. In addition, the spectre of Harold Shipman now haunts the euthanasia debate. For wasn't he able to get away with the 200 or so murders precisely because of an irresponsible lack of supervision within the medical profession and the law? Opponents of euthanasia will call for much tighter and more explicit legislation to prevent future Shipmans. For the perverse possibility exists that Shipman might even have thought that he was acting in his victims' best interests. Indeed, he might even have believed that he had obtained their explicit consent to that end. The answer, conclude the opponents, is more precise guidelines, tighter supervision, accountability and record-keeping.

Proponents of euthanasia, such as Dignity in Death (formerly the Voluntary Euthanasia Society), will agree to tightening up the net to prevent future Shipmans, but will argue for explicit criteria that patients would have to meet, in order to be granted their wish to die. In other words, euthanasia should be bound by the principles of the ideal contract: each side is free and informed about the present situation, about the available options and their consequences, and no undue enticement or coercion is involved. The purpose of the law is to regulate the contractual intercourse, enforce it if necessary, and prevent the violation of its terms, and all of this should be explicit and strict.

The problem with this criticism is that the law is often *more* than an instrument for regulation and enforcement; it is also a powerful symbol of the most important values in a society. On the one hand, the present law expresses an unequivocal message about the value of life – of *any* human life, even the most impaired – and about the seriousness of harming or taking it. Normally, the only plausible defence for killing another is in self-defence proportionate to the reasonably perceived threat, or in communal self-defence as part of a country's armed forces waging a just war. But this defence can hardly be invoked in the case of Tony Bland or Ms B. On the other hand, the law *implicitly* acknowledges that some situations can truly be so awful for a patient that it does make sense to say that they would be 'better off dead'. Cases of great suffering during a terminal illness, reinforced by direct entreaties by the patient, would seem the most likely case. But this acknowledgement has to be implicit, because, once again, the whole point of the principle of sanctity of life enshrined in the law would be lost if it were thought of as subject to explicit qualification in this way. To reiterate the important contrast the law is concerned with justification, with the giving and taking of reasons in a public space; whereas what is at stake in these extreme cases may be necessity.

Not only is there an ideal of full transparency and consistency in mainstream ethical theory, but there is in a common sense view of the law as well. This fudge seems to fly in the face of both. Once again, however, this is due to the distinctly opaque nature of ethical problems at the margins of life. In order to grasp this opacity and to see the limits of the transparency presumption, consider the responses of praise and blame. Normally we would praise a police officer for enforcing a speed limit, even if offered a bribe from the driver to neglect it. However, even if it were best that this patient should die, it would surely be perverse to *praise* the doctor for carrying it out. The act of euthanasia is a loss for everyone involved, an occasion for sadness even if not always for grief, a time to curse the gods for their inscrutable cruelty in making it come to this. The success of Ken Harrison at the end of *Whose Life Is It Anyway?* is hardly a triumph. In other words, the unqualified legal prohibition is important in helping to generate a climate of reluctance around the act of euthanasia: it must always be the very last option, even if it might be 'for the best'.

Recall the splendid Williams quotation about Agamemnon in Chapter 2. There he described how the notion of 'for the best' starts to lose its content in extreme situations. Similarly we may ask here: was it *really* the case that Ms B was better off dead? Who really knows? Not Ms B, not the doctors who eventually removed the ventilator. All we are left with is the force of Ms B's *will*, as if the whole legal rigmarole that surrounded her request was no more than a test of motivation, or a test of her anger. But whether this was for the better is surely impossible to say.

This is a bit limp by way of conclusion. But then again this book could be described as comprising little more than a series of insights knitted around a couple of more-or-less coherent threads; it is certainly not a systematic policy document. Much of the euthanasia debates I find confusing, and I have tried to elucidate some aspects of it; but I certainly do not claim to have solved any of the problems that will continue to beset us for many more years to come. The liberal experiments in Oregon and the Netherlands can certainly not be considered an unambiguous success, not only because it is too soon to tell but also because there are so many difficulties to interpreting the data with any confidence.

The only conclusion I can think of for a chapter like this, and indeed for a book like this, is a proposal to return to the paradigm of the long-term relationship between the patient and the General Practitioner. Despite Shipman, and despite the ever-improving technical prowess of the hospital consultants, there remains a very real place for a single individual who can get to know his patients well over the long term. Certainly he needs the expert knowledge of the human body and its mechanical defects. But more important is a rich life experience allowing him to identify and understand something of the many types of people who come to see him. Most of all, however, he needs the time and the will to develop relationships with his patients over many years. This will give him not only expertise in matters relating to technical medicine, but also a certain ethical authority in matters relating to life and death.

Those demanding greater transparency and consistency and accountability in public institutions, more detailed medical notes and clear application of rules will be disappointed by this paradigm and will worry about its potential for abuse. And some degree of institutionalisation of GPs might be necessary, although one allowing them as much flexibility as possible. For the margins of life will always remain stubbornly opaque, not only to managers and legislators, but also to the patients themselves. When medical technology can offer no further hope of cure, there is still the GP who can help because of his position of trust, knowledge and authority, and because of his access to the patient's home and to the patient's family. It is not an accident that many GPs say they feel privileged to have shared such final hours with their patients.

Notes

1 Technical Language and Ordinary Language

1. http://jme.bmj.com/misc/topten06.dtl (accessed 21 May 2007).
2. For most of this book I have tried to use the word 'ethical' rather than 'moral', since I am interested not just in the rules to guide conduct, but also in the ethical character behind that conduct – although the distinction is not crucial to my arguments. With this brief mention of Kant, I use 'moral', following his own use of it; and elsewhere I use 'moral status', following convention.
3. On the question of ethical expertise, see Noble (1982) and Weinstein (1994). My position is closer to that of Johnston (2001).

2 Ways of Seeing

1. It should not be thought that C experiences *no* ethical prohibitions, upon perceiving the chicken, of course – for example, he sees it as someone else's property that cannot be appropriated without the owner's informed consent. So even for C, it's not simply a *physical* chicken (two-legged, feathered etc.) but also an owned good, the appropriation of which would have to be appropriately justified. And like V, the reason why C does not just grab the chicken (he is hungry, after all) is simply that 'it belongs to someone else'.
2. The locus classicus of the modern debate is probably based on the following book: Singer (1975).
3. The phrase is Bernard Williams's. His first paper on the topic was 'Practical Necessity' (1981), which was slightly modified in a later paper entitled 'Moral Incapacity' in Making Sense of Humanity, Cambridge: CUP 1995.
4. There is a similar clause in section 38 of the Human Fertilisation and Embryology Act 1990. In addition, a health care professional may resign from a multidisciplinary team when that team comes to a decision to withdraw treatment (including artifical nutrition and hydration), with the intention of allowing the patient to die.
5. Notice that these arguments are not in favour of abortion as such; Savulescu is not trying to argue with a particular objecting doctor about the ethical permissibility of performing abortions; rather, he is arguing about the coherence of objecting, *given* the legal and social background of the society and organisation in which he has chosen to work. In addition, Savulescu is not denying the possibility of making exceptions; but the exceptions have to be justifiable with reference to the general principles that underlie the policy, such as the best interests or autonomy of a patient. For example, a surgeon can legally refuse to perform an operation if he is not convinced that the patient's consent is sufficiently full, free and informed.

6. This approach to the different kinds of why-questions comes from D. Z. Phillips (1992), who uses the example of the mother of a soldier killed in battle.

3 The Place of Pregnancy and Birth in Human Lives

1. See Hume's *An Enquiry Concerning Human Understanding*, section x.
2. Think of Ridley Scott's classic science fiction film *Alien*, where one of the (male) characters is orally 'impregnated' with an embryo that then grows and one day bursts out of his belly, killing him. The film's power draws directly from the experience of normal human impregnation, pregnancy and childbirth. For a fascinating discussion of this parallel, see Stephen Mulhall (2002).
3. This section owes a good deal to the insights of Carl Elliott (1992).
4. I am reminded here of the anatomical displays of Von Hagen in the travelling exhibit *Bodyworlds*. A number of human cadavers had been 'frozen' in a special plastic, flayed and partially dissected, and then positioned into rigid positions corresponding to recognisable human activities such as playing chess or swimming.

4 The Clash of Perspectives

1. http://www.opsi.gov.uk/acts/acts1990/Ukpga_19900037_en_1.htm (accessed on 21 May 2007).
2. The only exceptions would be those rare requests to a court, for a woman with a severe mental handicap to be forcefully implanted with a contraceptive device or even sterilised. Note that I am talking about preventing conception here, and thus have not mentioned those cases where children are removed by social services right after birth, on the grounds that the mother has a demonstrable lack of ability or interest in bringing up the child with sufficient care.
3. Children Act 1989; NHS and Community Care Act 1990; Adoption and Children Act 2002.
4. See, for example, the articles by Spriggs, Levy, and Anstey in the symposium of the *Journal of Medical Ethics*, vol. 28, 2002.
5. The actual case that sparked this controversy involved a congenitally deaf lesbian couple from Texas, who found a sperm donor who was himself congenitally deaf – so it was not a question of testing existing embryos for 'deafness'. See the previous note on *JME* symposium in n. 4 for details.
6. In a similar vein to Harris, Julian Savulescu proposes a 'Principle of Procreative Beneficence', according to which parents have a moral duty to select the embryo that is most likely to develop the best attributes (Savulescu 2001). Michael Parker (2007) criticised this in a similar way as I am criticising Harris, but with a different emphasis: 'Complex concepts, such as those of the good life, the best life, and human flourishing, are not reducible to simple elements or constituent parts which might be identified through the testing of embryos.'
7. I agree with Harris that it therefore makes no sense for an adult to sue his parents, or to sue an agency like the HFEA, for allowing him to be born, that is, to sue them for 'wrongful life'. At most he can submit a legitimate claim

against a public health system, for support on the basis of his *present* physical or psychological needs.

8. A late-onset detectable genetic condition, such as breast cancer or Huntingdon's Chorea, would be less 'severe' than Down's, I suggest, in that the child can expect at least three or four decades of normal life. As such it becomes even less of a reason to abort an afflicted foetus or embryo.

9. I am ignoring for the moment arguments that handicaps like Down's are at least partly socially constructed. It is true that our society could do more for people with Down's, and that this may reduce the number of abortions of Down's foetuses.

10. See, for example, John Rawls's (1971) famous thought experiment called 'The Original Position'.

11. http://www.bshg.org.uk/for_patients/for_patients.htm, (accessed on 21 May 2007).

12. Some detectable genetic conditions can be repaired, such as PKU (phenylketonuria). In addition, some genetic conditions are localised and delayed, and so there is time to do something with the areas likely to be affected. I am thinking in particular of the BRCA1 and BRCA2 genes, which mark a predisposition to breast cancer, and where the person can have a prophylactic bilateral mastectomy.

13. As in many places in this book, I am assuming the truth of Bernard Williams's controversial 'reasons internalism' ('Internal and external reasons', in Williams 1981). A reasons-externalist like Kant believes that there may be reasons *for* a given agent to do something, whether or not he (i) knows about the reason or (ii) accepts the reason. Williams argues, persuasively in my opinion, that the only genuinely external reasons have to do with the agent's inferential errors and ignorance of relevant facts. Otherwise there are no external reasons. For a reason can only genuinely function as a reason if it finds purchase – or could realistically find purchase – in what Williams calls the agent's 'subjective motivational set'.

5 The Abortion Debates

1. Appendix B 'Disposal following pregnancy loss before 24 weeks' gestation' to the Human Tissue Authority Code of Practice, dated 5 July 2006. (The Authority was established by the Human Tissue Act 2004.)

2. It is perhaps for this reason that infertility can be so devastating for women in a way that men cannot fully appreciate. It is tempting to ask why IVF should be subsidised by the NHS when the defect in question is hardly life threatening. But the point is that infertility can threaten the *meaning* of some women's lives by denying them the transformation to motherhood which they might consider their highest calling.

6 The Shape of a Life

1. It is true that the audience of films and television *is* passive, in the sense in question. I suggest, however, that these artforms are derivative of the essential storytelling mode of face-to-face conversation.

2. Taylor footnotes this with a reference to the work of Michael Bakhtin, especially his *Problems of Dostoyevsky's Poetics*.
3. In a similar vein, I have always been struck by the claim in the American Declaration of Independence that all men had a right to life, liberty and the *pursuit of happiness*. Surely pursuing happiness is the worst way to achieve it?

7   The Problem of Suicide

1. For a wide-ranging discussion of these arguments, together with discussions of rationality, see Battin 1996.
2. The text of the European judgement is available at: http://www.echr.coe.int/Eng/Press/2002/apr/Prettyjudepress.htm, (accessed on 21 May 2007).
3. In fact, the Crown Prosecution Service would have been unlikely to imprison Brian Pretty if he had carried it out. An article in the *Guardian*, on 20 October 2006 p. 5 describes five similar incidents where the defendants received suspended sentences or conditional discharges.
4. My thanks to Nicola Holtom, consultant palliative care specialist, for her comments on this section.

8   Making Sense of Dementia

1. Cited in Dworkin (1993), p. 220. The original reference is Firlik A., 'Margo's logo', in *Journal of the American Medical Association*, vol. 265, 1991.
2. My thanks to David Levy for a good deal of feedback on this chapter.
3. There is a similar problem in theological discussions about the resurrection of the body: with which of my many bodies do I pass into heaven? I would not be in much of a position to enjoy heaven with the last body I had while alive, aged 90 years, crippled and weak, and riddled with cancer.

9   Human Bodies

1. This chapter has been influenced by Annemarie Mol (2002).
2. The child will feel pain, certainly, but will not know that *this* awful sensation is 'pain', until he learns to ascribe the concept correctly to other people, and that means correctly recognising pain behaviour in response to a typically pain-causing situation, and recognising appropriate third-party responses (e.g. comforting) to the person in pain. For while a parent can define tables and chairs by pointing, there is no way he can point at an object in the child's experience to define the concept of pain. Importantly, it would not help at all were the parent to explain the neurophysiological mechanisms of pain from a textbook or by using a dissected frog. The best discussion of the concept of pain is of course in Wittgenstein's *Philosophical Investigations*.
3. In a Chapter 3 endnote I mentioned Von Hagen's anatomical exhibition *Bodyworlds*. This exhibit yielded the same sort of collision between the familiar and the uncanny, but the collision could have been much more striking and disturbing if he had not systematically flayed all the *faces*. Partly this was to preserve the anonymity of the donor, but partly it was to preserve the

more comfortable impression that these were *merely* bodies, and not people, on display.
4. *The Report of Royal Liverpool Children's Inquiry*, The Stationary Office 2001.
5. There is a separate question, which I shall not consider here, of whether the harm caused was sufficient to justify compensation by the NHS Trust involved, compensation that could be described as therefore removed from expenditures on patient care.
6. In the same journal issue, the pathologist H. Emson even writes that 'the body should be regarded as on loan to the individual from the biomass, to which the cadaver will inevitably return' (Emson 2003 p. 125). There should be no right to withhold organs at all, he argues. I find this image of a loan rather comical, as if the bodyless individual goes down to his local library to borrow a body for three score years and ten; and when his time is up, he just leaves the body.

10 The Euthanasia Debates

1. This can be confirmed by checking the government statistics at: http://www.statistics.gov.uk (accessed on 21 May 2007).
2. The case of Ms B is filed under *B (Ms)* v. *an NHS Trust* [2002] EWHC 429; [2002] 2 FCR 1. The case of Tony Bland is filed under *Airedale NHS Trust* v. *Bland* [1993] 1 All ER 821 HL.
3. This discussion was strongly influenced by McCall Smith's (1999) support for what he calls the 'middle ground' in the attitude of UK law to euthanasia.

Bibliography

Anscombe, E. 'Modern moral philosophy' [1958], in *Collected Philosophical Papers*, vol. III, Oxford: Blackwell 1981.

Battin, M. *The Death Debate: Ethical Issues in Suicide*, London: Prentice-Hall 1996.

Bayley, J. *Iris: A Memoir of Irish Murdoch*, London: Abacus 2002.

Beauchamp, T. and Childress J. *Principles of Biomedical Ethics* (1st ed.), Oxford: OUP 1979.

Brandt, R. 'The morality and rationality of suicide', in Perlin, S. (ed.), *A Handbook for the Study of Suicide*, Oxford: OUP 1975. Reprinted as 'The rationality of suicide', in Battin, M. and Mayo, D. (eds), *Suicide: The Philosophical Issues*, New York: St. Martin's Press 1980.

Brink, D. *Moral Realism and the Foundations of Ethics*, Cambridge: CUP 1989.

Brody, H. *Stories of Sickness* (2nd ed.), Oxford: OUP 2003.

Buchanan, A. 'Advance directives and the personal identity problem', in *Philosophy and Public Affairs*, vol. 17, no. 4, 277–302, Autumn, 1988.

Callahan, D. 'Aging and the goals of medicine', in *Hastings Center Report*, vol. 24, no. 5, 39–41, 1994.

Cavell, S. *The Claim of Reason*, Oxford: OUP 1979.

Charon, R. and Montello, M. *Stories Matter: The Role of Narrative in Medical Ethics*, London: Routledge 2002.

Clark, B. *Whose Life Is It Anyway?* Speakerman (ed.), London: Heinemann International 1989.

Curran, C. 'Abortion: Contemporary debate in philosophical and religious ethics', in Reich, E. (ed.), *Encyclopaedia of Bioethics*, New York, Free Press 1978.

Dewar, S. and Boddington, P. 'Returning to the Alder Hey report and its reporting: Addressing confusions and improving inquiries', in *Journal of Medical Ethics*, vol. 30, 463–9, 2004.

Diamond, C. *The Realistic Spirit*, Boston: MIT Press 1991.

Dworkin, R. *Life's Dominion*, New York: Alfred A. Knopf 1993.

Elliott, C. 'Where ethics comes from and what to do about it', in *Hastings Center Report*, vol. 22, 28–35, 1992.

Elliott, C. *A Philosophical Disease: Bioethics, Culture and Identity*, London: Routledge 1999.

Emson, H. 'It is immoral to require consent for cadaver organ donation', in *Journal of Medical Ethics*, vol. 29, 125–7, 2003.

Finnis, J. 'Misunderstanding the case against euthanasia', in Keown, J. (ed.), *Euthanasia Examined: Ethical, Clinical and Legal Perspectives*, Cambridge: CUP 1995.

Gaita, R. *A Common Humanity: Thinking about Love and Truth and Justice*, London: Routledge 2000.

Gaita, R. *Good and Evil: An Absolute Conception* (2nd ed.), Routledge 2004.

Gillon, R. 'Ethics needs principles – four can encompass the rest – and respect for autonomy should be "first among equals"', *Journal of Medical Ethics*, vol. 29, 307–12, 2003.

Glover, J. *Causing Death and Saving Lives*, London: Penguin 1977.

Goldie, P. 'Time and Narrative', unpublished manuscript, 2001.

Harris, J. 'One principle and three fallacies of disability studies', in *Journal of Medical Ethics*, vol. 27, 383–7, 2001.

Harris, J. 'In praise of unprincipled ethics', *Journal of Medical Ethics*, vol. 29, 303–6, 2003.

Hope, T., Savulescu, J. and Hendrick, J. *Medical Ethics and Law: The Core Curriculum*, London: Churchill Livingstone 2003.

Hursthouse, R. 'Virtue theory and abortion', in *Philosophy and Public Affairs*, vol. 20, no. 3, 223–46, Summer 1991.

Jackson, J. *Ethics in Medicine*, London: Polity 2006.

James, W. 'The will to believe', in *The Will to Believe and Other Essays in Popular Philosophy*, New York: Dover 1952.

Johnston, P. 'Bioethics, wisdom and expertise', in Elliott, C. (ed.), *Slow Cures and Bad Philosophers: Essays on Wittgenstein, Medicine and Bioethics*, Durham, NC: Duke UP 2001.

Jonsen, A. 'Commentary: Scofield as Socrates', in *Cambridge Quarterly of Healthcare Ethics*, vol. 2, no. 4, 434–8, 1993.

Joyce, J. *Dubliners*, London: Penguin Popular Classics 1996.

Keown, J. 'The case of Ms. B: Suicide's slippery slope?' in *Journal of Medical Ethics*, vol. 28, 238–9, 2002.

Keown, J. *Euthanasia, Ethics and Public Policy: An Argument Against Legislation*, Cambridge: CUP 2002.

MacIntyre, A. *After Virtue* (2nd ed.), London: Duckworth 1985.

Mackie, J. *Ethics: Inventing Right and Wrong*, London: Penguin 1977.

Maclean, A. *The Elimination of Morality: Reflections on Utilitarianism and Bioethics*, London: Routledge 1993.

Marquis, D. 'Why abortion is immoral', in *Journal of Philosophy*, vol. 86, no. 4, 183–202, 1989.

McCall Smith, A. 'Euthanasia: The strengths of the middle ground', in *Medical Law Review*, vol. 7, 194–207, Summer 1999.

McMahon, J. *The Ethics of Killing: Problems at the Margins of Life*, Oxford: OUP 2002.

Mol, A. *The Body Multiple: Ontology in Medical Practice*, Durham, NC: Duke UP 2002

Mulhall, S. 'Fearful thoughts', a review of J. McMahon's *The Ethics of Killing*, in *London Review of Books*, vol. 24, no. 16, 22 August 2002, http://www.lrb.co.uk/.

Mulhall, S. *On Film*, London: Routledge 2002.

Nelson, H. *Stories and Their Limits: Narrative Approaches to Bioethics*, London: Routledge 1998.

Noble, C. 'Ethics and experts', in *Hastings Center Report*, vol. 12, no. 3, 7–15, 1982.

Parfit, D. *Reasons and Persons*, Oxford: OUP 1984.

Parker, M. 'The best possible child', in *Journal of Medical Ethics*, vol. 33, 279–83, 2007.

Phillips, D. Z. 'Allegiance and change in morality', in *Interventions in Ethics*, London: Macmillan 1992.

Phillips, D. Z. 'In search of the moral "must"', in his *Interventions in Ethics*, London: Macmillan 1992.

Rawls, J. *Theory of Justice*, Cambridge, MA: Harvard UP 1971.

Rhees, R. *Moral Questions*, Phillips D. Z. (ed.), London: Macmillan 1999.

Robertson, J. *Children of Choice: Freedom and the New Reproductive Technologies*, Princeton, NJ: Princeton UP 1993.

Savulescu, J. 'Procreative beneficence: Why we should select the best children', in *Bioethics*, vol. 15, 413–26, 2001.

Savulescu, J. 'Death, us and our bodies: Personal reflections', in *Journal of Medical Ethics*, vol. 29, 127–30, 2003.

Savulescu, J. 'Conscientious objection in medicine', in *British Medical Journal*, vol. 332, 294–7, 4 February 2006.

Singer, P. *Animal Liberation: A New Ethics for our Treatment of Animals*, New York: New York Review/Random House 1975.

Singer, P. 'Ms B and Diane Pretty: A commentary', *The Journal of Medical Ethics*, vol. 28, 234–5, 2002.

Steinbock, B. 'Preimplantation genetic diagnosis', in Burley, J. and Harris, J. (eds), *A Companion to Genethics*, Oxford: Blackwell 2002.

Strawson, P. 'Freedom and resentment', in *Freedom and Resentment and Other Essays*, London: Methuen 1974.

Taylor, C. *The Ethics of Authenticity*, Toronto: University of Toronto Press 1991.

Teichmann, R. 'The functionalist's inner state', in Schroeder, S. (ed.), *Wittgenstein and Contemporary Philosophy of Mind*, London: Palgrave Macmillan 2001.

Thomson, J. 'A defence of abortion', in *Philosophy and Public Affairs*, vol. 1, no. 1, 47–66, 1971.

Velleman, D. 'Against the right to die', in *Journal of Medicine and Philosophy*, vol. 17, no. 6, 665–81, 1992.

Warnock, M. *A Question of Life*, London: Blackwell 1985.

Warnock, M. *Making Babies: Is There a Right to Have Children?* Oxford: OUP 2002.

Warren, M. 'On the moral and legal status of abortion', in *The Monist*, vol. 57, no. 1, 665–81, 1973.

Weinstein, B. 'The possibility of ethical expertise', in *Theoretical Medicine and Bioethics*, 15, 61–75, 1994.

Wilkinson, S. *Bodies for Sale: Ethics and Exploitation in the Human Body Trade*, London: Routledge 2003.

Williams, B. 'Ethical consistency', in *Problems of the Self*, Cambridge: CUP 1973.

Williams, B. *Moral Luck*, Cambridge: CUP 1981.

Williams, B. *Ethics and the Limits of Philosophy*, Cambridge, MA: Harvard UP 1985.

Winch, P. 'Moral integrity', in his *Ethics and Action*, London: Macmillan 1972.

Wyatt, J. 'Medical paternalism and the foetus', in *Journal of Medical Ethics*, vol. 27, suppl. II, 15–20, 2001.

Index